The Disorder of Things

THE DISORDER OF THINGS

Metaphysical Foundations
of the Disunity of Science

JOHN DUPRÉ

Harvard University Press
Cambridge, Massachusetts, and London, England 1993

Library of Congress Cataloging-in-Publication Data

Dupré, John.
   The disorder of things: metaphysical foundations of the disunity of science /
John Dupré.
     p. cm.
   Includes bibliographical references and index.
   ISBN 0-674-21260-6
   1. Chaotic behavior in systems. 2. Reductionism. 3. Determinism (Philoso-
phy) I. Title.
Q172.5.C45D87 1993
501—dc20
92-12930
   CIP

For my mother, Catherine Dupré,
and in memory of my father, Desmond Dupré

Acknowledgments

The investigations out of which this book developed have extended over some dozen years. Consequently the advice, suggestions, and criticisms of many teachers, students, colleagues, and friends have become inextricably accreted into the text. It would be impossible to acknowledge all these debts, if only because I no longer remember what many of them are. Therefore, I shall attempt only to indicate the largest and most obvious.

I would like to thank Peter Hacker, from whose philosophical expertise, knowledge, and encouragement I have benefited for more than twenty years as student, colleague, and friend. A large number of friends and colleagues at Stanford University, both in and outside the Department of Philosophy, have helped to provide a working environment over the last decade as intellectually stimulating as it is socially congenial. I must especially mention my former colleague Nancy Cartwright, in discussion and collaboration with whom over many years my views on causality, in particular, have developed; and John Perry, who has provided uniquely penetrating criticism and generous encouragement of my work, and support of my philosophical career, for as long as there have been such things to criticize or support. In the university at large, a number of colleagues in the Feminist Studies Program and the Cultural Studies Group, in addition to sharing their knowledge and insight, have constantly reminded me both that there are many different approaches to fundamental issues, and that the importance of such issues is much more than "academic."

Like most people who have worked in the philosophy of biology, I have received generous and invaluable help and encouragement from David Hull. And Philip Kitcher has provided a quantity of criticism, friendship, and support—including detailed comments on a late version of the entire manuscript of this book, without which, whatever the merits of this work, they would have been less. As will be apparent from the references, the published works of these two philosophers have been crucial in forming my understanding of the philosophy of biology.

At widely different times I have received generous and helpful comments from various other people on recent or ancestral versions of parts of this work. I am especially grateful to Michael Bratman, Richard Burian, Barbara Horan, Nick Jardine, Hugh Mellor, and the members of a graduate seminar at Stanford in 1990 in which the main ideas of the book were discussed.

I am indebted for financial support to the Pew Memorial Trust; a grant from the Systems Development Foundation to the Center for the Study of Language and Information, Stanford; and most especially to the Stanford Humanities Center, where I was privileged to spend a year during an early phase of this work in what must surely be the perfect surroundings for scholarly research.

My greatest debt, finally, is to Regenia Gagnier. I cannot quantify the benefits that I have received from her extensive learning, broad vision, and detailed criticism. Without her help and intellectual stimulation it is doubtful whether the book would have been written at all.

Contents

The human understanding is of its own nature prone to suppose the existence of more order and regularity in the world than it finds. And though there may be many things in nature which are singular and unmatched, yet it devises for them parallels and conjugates and relatives which do not exist.

Francis Bacon, *The New Organon* (1620), XLV

Introduction

Science and Metaphysics

This book has two interwoven theses. The first concerns science. It is the denial that science constitutes, or could ever come to constitute, a single, unified project. The second is metaphysical, a thesis about how the world is. This thesis is an assertion of the extreme diversity of the contents of the world. There are countless kinds of things, I maintain, subject each to its own characteristic behavior and interactions. In addition, I propose a relation between these two theses: the second shows the inevitability of the first.

As my topics are the philosophy of science and metaphysics, so my argument will draw eclectically from science, philosophy, and common experience. Despite some skeptical notes about parts of science, I place myself firmly in the philosophical tradition that sees empirical, often scientific, inquiry as providing the most credible source of knowledge of how things are. In contrast to most related endeavors, however, I shall draw primarily not on physics, but on biology. Biology is surely the science that addresses much of what is of greatest concern to us biological beings, and if it cannot serve as a paradigm for science, then science is a far less interesting undertaking than is generally supposed.

It is now widely understood that science itself cannot progress without powerful assumptions about the world it is trying to investigate, without, that is to say, a prior metaphysics. Aristotle's picture of the natural ordering of basic substances in concentric circles and Newton's vision of a universe of massive objects moving through an infinite void were not the products of empirical inquiry, but sets of assumptions that proved among the most fruitful in history in suggesting strategies of investigation and in interpreting the results of those investigations. Such assumptions, concerning such matters as the unity or diversity of the world's ultimate contents, or the nature and prevalence of causality, are the kinds of questions I take to be in the domain of metaphysics. The metaphysics of modern science, as also of much of modern Western philosophy, has generally been taken to posit a deterministic, fully law-governed, and potentially fully intelligible structure that pervades the material universe. The rejection of this set of assumptions is what I mean by "The Disorder of Things."

Given the dependence of science on metaphysics, it might be thought that the examination of science for metaphysical insights constitutes a *petitio principii*. The reason I deny this conclusion is that whereas I accept the dependence of scientific inquiry on a complex body of fundamental presuppositions, I also claim that empirical inquiry (which I do not limit to scientific inquiry) provides the evidence on which such assumptions must ultimately rest. Thus I claim that founding metaphysical assumptions of modern Western science, most notably those that contribute to the picture of a profoundly orderly universe, have been shown, in large part by the results of that very science, to be untenable.[1] And this, in turn, shows the impossibility of a unified science. I also want to argue, in the final part of the book, that these conclusions matter, that there are political as well as philosophical issues at stake in these questions about order and unity.

The Refutation of Essentialism, Reductionism, and Determinism

The best way to introduce the position against which this book will argue is to consider the continuing influence of one particularly notorious founding metaphor of modern science, the idea that the universe should be considered as a gigantic machine. Traditionally the favored machine has been a clock. Although the metaphor is naturally associated with a clockmaker, as in William Paley's famous argument for the existence of God, latter-day mechanists have deprived the clockmaker of no more than his sight.[2] Anyone who thinks that such mechanical metaphors have faded in significance might reflect on the amount of scientific effort that has been devoted in recent years to investigation of the hypothesis that the human brain—and generally also the human mind—is really a kind of calculating machine, or computer.[3]

The philosophical thesis most intimately connected with this mechanistic metaphor is determinism. The cosmic clock, we must assume given its provenance, has always told the correct time and always will. To achieve such precision its components must exhibit the same unvarying reliability as the whole. Once the clock is wound up and set in motion, its behavior and that of all its parts are determined for all eternity.

It is true that the contemporary cosmic clock has become complex beyond any possibility of full human comprehension. Indeed the current vogue for chaos theory suggests that certain aspects of this complexity may be in principle beyond the reach of certain kinds of comprehension, notably prediction. Nevertheless, the metaphysical conceptions of order that originated in the picture of the mechanical universe seem to have been little threatened, if sometimes modified, by such developments. The case of chaos theory nicely illustrates this resilience. Prediction, though long conceived as a very central excellence of scientific understanding, is a goal that has tended to recede rather than approach as various scientific disciplines have increased their understanding of the complexity of the phenomena within their domains. Chaos theory appears to confront prediction not merely with an insuperable practical difficulty, but with a logically impassable obstacle. Yet paradoxically determinism, the metaphysical underlay of the possibility of prediction, is strengthened rather than threatened by this development. For the central mathematical functions of chaos theory are quite deterministic. They serve to show that it was a mistake to assume that, in a deterministic universe, even Laplace's demon could predict the evolution of events. Even the increasing prevalence of probabilistic rather than deterministic hypotheses and methods from quantum mechanics up through the scientific hierarchy has not, I shall argue, led to the rejection of some of the most fundamental features of the deterministic world view.

Arguments against determinism, however, come at the end of

my exposition, approached by way of a rather more subtle aspect of the mechanistic paradigm, reductionism. The way to understand the behavior of a machine is first to understand the behavior of its component parts, and then to see how the interactions of those parts generate the behavior characteristic of the whole. For a complex machine this task may need to be carried out in a number of stages. Thus we might understand a car, in the first instance, as consisting of engine, gearbox, steering assembly, starter, brake system, and so on. Each of these would be subject to further structural analysis, and after a very few steps we would reach relatively simple parts with simply intelligible mechanical properties.

This picture of successive understanding in terms of a hierarchy of ever smaller and simpler structural components is generally referred to as reductionism and has been a dominant ideal of modern science. Reductionistic accounts of science can be developed in various ways. My elementary illustration suggests a hierarchy of kinds of objects. A more familiar philosophical interpretation of reductionism involves a hierarchy of theories, each of which is, ideally, derivable from a theory of simpler entities and some postulated identities between the entities of the first theory and structures of entities of the second theory. This position is often referred to as theory reductionism. It will be apparent that such a thesis suggests a strongly ordered and global structure to the universe. Theory reductionism seems to imply that in principle all our understanding of everything should be derivable from our understanding of the smallest structural components of the universe (assuming, at least, that any such smallest components exist). For this reason, since this would amount to the subsumption of all of science by the science of the microphysical, theory reductionism has become widely identified with the unity of science. The thesis also has strong affinities and connections with determinism. In particular, a deterministic account of the smallest constituents of things, classically a mechanistic story about minute and elastic atoms, will transmit its deterministic properties to the structures composed of those things.

One final philosophical doctrine that I identify as central to the mechanistic world view is essentialism. In contemporary philosophy this is most naturally approached through theories of natural kinds. According to such theories, one fundamental kind of facts, over and above facts about what individual things there are, are those concerning what *kinds* of things the world contains. To say that iron or humans form natural kinds is to claim a kind of objective legitimation for our divisions of things into those that are and are not ferrous or human. This idea is also related to mechanism. The fact that something is a car more or less determines what its constituent parts are and to what kinds they belong. Pistons and sparkplugs are proper parts of cars in a way in which that part of the engine block that still has its original paint, say, is not. Even more diffuse things can be distinguished, such as the cooling system. The radiator and the water pump are objectively part of the same constituent of the car in a way that the distributor cap and the cigarette lighter are not. The parts of cars form a determinate hierarchy of kinds. If the world is a machine, it is reasonable to suppose that the parts of the world do likewise.

In contrast with both determinism and reductionism, I do not doubt that a weak version of the doctrine of natural kinds is defensible. That there are objective divisions between some distinct kinds of things would be hard to deny. Indeed, I shall argue that there are many more such divisions than is generally allowed. What I do want to deny, and what is suggested by the embedding of an account of natural kinds in the context of mechanism, is the idea that it is a generally appropriate and tractable question to ask, of an object, What is the natural kind to which it belongs? I claim, on the contrary, that such questions can be answered only in relation to some specification of the goal underlying the intent to classify the object. In relation to an account of the workings of a car, it is quite straightforwardly true that a particular constituent object should be classified, for example, as a piston. This classification correctly identifies the (sole) function of this constituent in the overall economy of the larger system the car. But the unambiguousness of this identification is clearly relative to a particular context, the context defined by the overall function of the car. And it is quite possible for the career of the piston to exceed its tenure in the car and to continue as a hammer or a weapon. Since the world is not a machine, nature does not generally provide contexts that can serve to determine unambiguously the kinds to which objects belong, and such context must typically be provided instead by the goals of a particular investigation.

The idea that things belong to unambiguously discoverable natural kinds is intimately connected with the commitment to essentialism. Essentialism connects with conceptions of natural kinds through the idea that what makes a thing a member of a particular natural kind is that it possesses a certain essential property, a property both necessary and sufficient for a thing to belong to that kind. The essential property is thus admirably suited to provide an objective feature which can answer the question, independently of any context of inquiry, To what kind does this thing belong?

These various issues concerning classification are ideally addressed in relation to biology. Classification has always been a central task of biology, and biological classification is a significant part even of prescientific language. As I shall argue in Chapters 1 and 2, biological classification provides no encouragement to essentialism. Not only does essentialism fail in biology, but it can be argued that there is not even a unique set of kinds into which biological organisms should ideally be sorted.

The main body of this book will be occupied with detailed critiques of these three doctrines—determinism, reductionism, and essentialism—which constitute the central pillars of a classical conception of cosmic order. At the same time, I shall develop some ideas about what a universe lacking the order implied by these doctrines must be like. Partly to develop more perspicaciously the positive views, and partly because it seems the clearest way to present the argument, I shall discuss them in the reverse order of that in which I have introduced them here. Although my decision to begin with a discussion of kinds and essentialism might seem eccentric to a reader focused on the history of discussions of scientific unity, my views on these topics provide the natural starting point for explaining the alternative to the classically ordered universe that I propose. The most general positive doctrine I shall advocate is pluralism: first, in opposition to an essentialist doctrine of natural kinds, pluralism as the claim that there are many equally legitimate ways of dividing the world into kinds, a doctrine I refer to as "promiscuous realism";[4] and second, in opposition to reductionism, pluralism as the insistence on the equal reality and causal efficacy of objects both large and small. These pluralisms make determinism almost impossible to sustain. Only a privileged and restricted set of entities and kinds could make it plausible that everything could occur in accordance with a unified and universally applicable set of principles. I shall claim, on the contrary, that in the pluralistic universe that I take to be ours, causality is a much more scattered and variable feature.

The Disunity of Science

The idea of science as a project that might ultimately be completed in some grand synthesis of all natural knowledge is an understandable and perennial dream. The greatest scientist of the twentieth century, Albert Einstein, was a notorious believer in unification and simplicity as fundamental goals of science. He is quoted, for instance, as remarking: "What really interests me is whether God had any choice in the creation of the world."[5] Some of Einstein's greatest achievements have been attributed to his search for unifying principles; on the other hand, much of his later life was devoted to what is now seen as a fundamentally misconceived quest for a unified field theory. At any rate, if the world is as I have just briefly described it, then the dream of an ultimate and unified science is a mere pipe dream. A number of philosophers, perhaps a majority, have become skeptical of strong doctrines of scientific unity in recent years. But this skepticism has generally derived not from doubts about the traditional metaphysical underpinnings of a possible unified science, but from the recognition of insuperable pragmatic obstacles to its development: the cosmic clock has proved too complex for complete analysis by humans. Apart from endorsing and defending this skepticism, my goal is to anchor that skepticism in a more appropriate metaphysics. Thus my thesis will be that the disunity of science is not merely an unfortunate consequence of our limited computational or other cognitive capacities, but rather reflects accurately the underlying ontological complexity of the world, the disorder of things.

Concerns with the unity of science have a long history. The

expression "unity of science" is most widely associated with the Vienna circle and with logical positivism, and the epistemological monopoly of science that it is one of my major goals to criticize here was perhaps most unproblematically assumed by that philosophical school. However, for a number of reasons I shall not attempt to trace the issues I address to their roots in that source. First, and perhaps most fundamentally, a central defining characteristic of logical positivism was the rejection of metaphysics. This book, as its title declares, is intended to be metaphysical through and through. Second, the resurgence of metaphysics can be seen in the transmutation of one of my central topics, reductionism, the doctrine most generally associated with the unity of science. For classical positivism, reductionism refers to the attempt to ground knowledge entirely in the simplest possible observation statements, so as to eliminate any dubious inferences beyond the observable. This is the project of Rudolf Carnap's *Aufbau* (1928), the project perhaps best known from W. V. O. Quine's assault on it in his classic paper "Two Dogmas of Empiricism" (1951). The conception of unified science that derives from this project has now been long abandoned. It is, at any rate, far removed from the reduction of scientific theories to theories of the microphysical, which is the topic of a substantial proportion of this book. Whereas positivist reduction is strictly and intentionally epistemological, contemporary theory reduction, presupposing a structural hierarchy down to microphysics, is thoroughly ontological. Indeed, the microphysical, the basis for contemporary reductionism, was the primary object of positivist epistemological suspicion, and the main target of reduction to the observable.

The distance of contemporary ideas on the unity of science from those of the Vienna circle is even clearer in the light of recent work by Jordi Cat, Hasok Chang, and Nancy Cartwright (1991, 1992) on Neurath's conception of unified science. Otto Neurath, generally considered as the leader of the unity of science movement, appears to have been explicitly opposed to any kind of physicalist *or* sensationalist reduction.[6] Indeed Cat, Chang, and Cartwright (forthcoming) suggest that he held a pluralistic attitude to science very similar to the one that I defend. For Neurath, the point of the unity of science was just that particular practical problems may require input from many different parts of science for their solution, and all the parts of science should therefore be jointly applicable to the same question. The *Encyclopaedia of Unified Science* was not to be a systematic text, but a toolkit of scientific devices. The importance of this point of view, in turn, can be traced to Neurath's enthusiasm for, and practical involvement with, attempts to establish centralized socialist states, both as a participant in the unsuccessful socialist uprising in Bavaria in 1919, and subsequently as connected with the socialist government of Vienna from 1919 to 1934 (Cat, Chang, and Cartwright, forth-coming).[7]

For all of these reasons, for one, like myself, concerned with contemporary debates about the unity of science, the texts of classical positivism are of only marginal relevance. There are, of course, important threads that could be followed from the work of Carnap, Moritz Schlick, Neurath, and others to present-day concerns. But such historical exploration is not among my aims. While the ghosts of innumerable ancestors undoubtedly walk through any philosophical text, the ancestors I shall directly en-gage are recent ones.[8] Quite specifically, the starting points for the conversations in which I see myself as engaged are mostly around the 1950s and early 1960s. This post-positivist but pre-Kuhnian watershed in the philosophy of science, marked by classics such as Ernest Nagel's *The Structure of Science* (1961) (the benchmark for contemporary discussions of reductionism) and C. G. Hem-pel's *Aspects of Scientific Explanation* (1965), represent the high point of sophisticated and shared optimism about the rationality, coher-ence, and value of science.

Since then, in substantial part as a result of the influence of Thomas Kuhn, things have fallen apart. My present concerns are with various fragments of this disintegration. One set of fragments comprises the thriving areas of philosophy of science that have avoided questions about science in general by focusing on prob-lems peculiar to specific scientific disciplines. Especially prominent are the philosophy of physics and the philosophy of biology; several of the following chapters engage mainly with work that has come out of the latter field. In a more general way, the philosophical views of many of the leading philosophers of the last several decades reflect the influence of both positivist and post-positivist philosophies of science. Quine, one of the most influential critics of positivism, retains an exclusive role for science in ontology that I would consider insupportable. Donald Davidson's views in the philosophy of mind, though explicitly intended as antireductionistic, rest, I argue in Chapter 7, on a metaphysics that is entirely and unacceptably reductionistic and deterministic. In the philosophy of science aspects of post-positivist or even positivist views of science can be found everywhere, though generally not united in the works of any one philosopher. Many philosophers of science, indeed, still adhere to some doctrine of scientific unity, though often doctrines rather different from those I have mentioned, proposing, for example, unity of method or of process. Some of the most important such doctrines will be considered in Chapter 10. One of my aims, then, is to clear away some of the remaining debris of a philosophical movement that, though largely abandoned some decades ago, continues to exercise a powerful influence over major areas of philosophy.

I shall, finally, offer some constructive if sketchy suggestions about how to think of science in the absence of unity. Given that there may be little common to all the various activities or systems of belief that coexist under the conceptual umbrella of science, I shall propose that science is best seen, in Wittgenstein's valuable phrase, as a family resemblance concept. Many features are common to many sciences, but no set is definitive of any adequate science. This will be recognized as an application to science of the antiessentialism that I shall defend for nature. In accordance with this parallel, and rejecting the possibility of any single criterion that constitutes a body of belief as scientific and hence epistemically acceptable, I also advocate epistemological pluralism. While I am reluctant to advocate the extreme tolerance of Paul Feyerabend (1975)—I have stronger prejudices against astrology, theology, and even alchemy—I would certainly reject the dogmatic monotheism of much contemporary philosophy of science: there are surely paths to knowledge very different from those currently sanctioned by the leading scientific academies. In pursuit of this idea, I would like to suggest that rather than seeking a criterion of scientificity, we should attempt to develop a catalogue of ep-

istemic virtues. Some of these will flow naturally from the philosophical tradition: empirical accountability, consistency with common sense and other well-grounded scientific belief, and perhaps
the more aesthetic virtues such as elegance and simplicity. More
recent conceptions of science suggest that these must be supplemented with more straightforwardly normative virtues. Investigation of the androcentric and ethnocentric biases of much science
suggests, for example, a fundamental desideratum of democratic
inclusion and accountability.[9]

Critical Perspectives on Science

As can be inferred from my skepticism about the unified project
of science, my goal is not to provide a general criticism of the
aspirations or achievements of science. For such a project presupposes that some entity—science—exists to be identified and criticized. And while there may perhaps be a genuine *sociological* object
identifiable as the referent of "science," it is obvious that this
would be quite inadequate to capture the concept assumed in most
philosophy of science both traditional and contemporary. To mention only the most obvious point, science is generally taken to
have some more or less privileged claim to truth. Yet there are
large areas of science in the sociological sense that, at least to
the casual glance, have at best highly questionable claims to truth
or credibility. One reason why it is important to dispose of the
myth of scientific unity is that one might, in principle, judge that
macroeconomics or mathematical population genetics, say, has
claims to credibility on a par with palmistry or tarot reading,
without being committed to making the same claim about mechanics or immunology.

Skepticism about cosmic order should raise doubts not only
about the unity of science but also about its universality. Since
science does, presumably, presuppose some kind of preexisting
order in the phenomena it attempts to describe, limits to the
prevalence of order may entail limits to the applicability of science.
Some areas of science may fail because the subject matter is in-
hospitable to scientific methods. A particularly interesting possibility of this kind is that there might be large parts of the human

sciences in which patterns do not exist to be discovered, but are rather created by our decisions to conceive of them in particular ways. This possibility would show that scientific theories in such an area could not be merely descriptive, but must be at least covertly normative.

The separation of the scientific sheep from the goats should not be seen merely as a claim that some theories are not true. Such a claim will hardly seem surprising to students of contemporary science studies. Many theorists, for varying reasons, hold that no scientific theories are true. Many take the familiar pessimistic induction on the history of science (all past theories have turned out false, so probably ours will too) as sufficient to establish this thesis. In philosophy, particularly as a result of the influence of Bas van Fraassen (1980), it has been fashionable recently to deny not only that scientific theories are true, but even that the entities to which they purport to refer exist.[10] This antirealist position claims that all we should be concerned with in evaluating scientific theories is the extent to which the predictions they make about observable things are true. The sociology of knowledge movement, for quite different reasons, has advocated a consistent repudiation of the role of truth in any aspect of the explanation of scientific belief. I wish to distinguish my position sharply from each of those just mentioned. One reason that I need to do so is, paradoxically perhaps, that both of these positions tend to be ultimately conservative.[11] This is because both projects, being quite unrestricted in their scope, provide no motivation for questioning the specific epistemic credentials of particular scientific projects. That is, they typically assume that the domain to which their analyses apply is given in advance, tacitly presupposing some kind of scientific unity.

This conservatism is clear, if somewhat surprising, in the case of the sociology of knowledge movement. By asserting that all scientific belief should be explained in terms of the goals, interests, and prejudices of the scientist, and denying any role whatever for the recalcitrance of nature, it leaves no space for the criticism of specific scientific beliefs on the grounds that they do reflect such prejudices rather than being plausibly grounded in fact. The uncongeniality of the sociology of science program to thinkers genuinely concerned with political influences on scientific belief is nicely stated by the feminist philosopher of science Alison Wylie (1992): "Only the most powerful, the most successful in achieving control over their world, could imagine that the world can be constructed as they choose." None of this is to deny the importance of what investigations by participants in this program have contributed to our understanding of the role of social, political, and personal factors in the formation of scientific belief. I want only to insist that these must be understood in interaction with some account of justifiable belief.

Although realism is an important part of my position—I believe that there are many things in heaven and earth, not merely in the human mind—the debate between realists and antirealists is somewhat tangential to my concerns. Neither side is primarily concerned with making distinctions among scientific theories, so the issue just raised of conservatism is moot. Indeed it seems almost a dogma of contemporary philosophy of science that evaluation of particular scientific theories is a task to be left to the genuine scientific priesthood rather than to their philosophical acolytes. In some ways antirealism may seem a pleasant relief from the baleful dominant tendency in philosophy of science, a reverence for the products of science verging often on the obsequious, and leading sometimes to the desire to annex epistemology, metaphysics, and even ethics in the name of some trendy "scientific" adventure.[12]

One reason for favoring realism, on the other hand, should be mentioned. The only way to put the genuine insights of the sociology of knowledge program, not to mention a good deal of more orthodox history of science, into a proper perspective is to see the kinds of forces they describe as interacting with a real and sometimes recalcitrant world. In fact, I believe that these aspects of science are inextricably intertwined, so that "good" science involves inseparable elements of the epistemically good and the sociopolitically good. My overall thesis is that there is a lot of good science and a lot of bad. Neither the uniform reduction of science to the nonepistemic, nor the concession of all epistemic matters to those who choose to call themselves scientists, can adequately pursue this insight.

In the final chapter I shall take up the question of the social and political aspects of science, giving some attention to the crucial insight that science is, indeed, a human practice serving identifiable human social and individual goals. (This perspective is, however, foreshadowed in my arguments in Part I that classifications often, at least, require a humanly determined goal.) I do not suppose that pursuit of this insight can provide a *sufficient* account of science, but I am confident that no account that ignores it can be adequate. A great deal of work has been done in recent years on the role of political and social influences in the development of science, and on political and social biases in the products of science. Of particular importance to my understanding of this topic is the contribution of a number of feminist critics of science.

I cannot attempt a comprehensive discussion of these issues, which would require another book. What I do hope to show is that the conclusions reached in the earlier chapters have important bearing on issues concerning science and value. First, the denial of scientific unity defuses any suggestion that the rejection of substantial areas of science as mere ideology calls into question the entire scientific edifice of which the former are alleged to be integral parts. Second, the denial of the universality of science raises serious questions whether *any* prima facie scientific investigation is the appropriate way to approach certain domains of inquiry. Putatively scientific accounts of how things objectively are can profoundly obfuscate debates in areas such as economics, where the real issue is how things should be made to be. Skepticism about the very existence of a preexisting *order* in such domains should make this view more plausible. Finally, my discussion of causality and defense of indeterminism lead to an unorthodox defense of the traditional doctrine of freedom of the will. Very simply, the rejection of omnipresent causal order allows one to see that what is unique about humans is not their tendency to contravene an otherwise unvarying causal order, but rather their capacity to impose order on areas of the world where none previously existed. In domains where human decisions are a primary causal factor, I suggest, normative discussions of what ought to be must be given priority over claims about what nature has decreed.

I. Natural Kinds and Essentialism

1. Natural Kinds

Classification

Distinctions are easy enough to draw. Useful ones are another matter. It is often supposed that classification, and especially scientific classification, the distinguishing of the different kinds of things in the world, is an activity that has revealed, or can be expected to reveal, an orderly, unique, and perhaps hierarchical arrangement of the things that exist. For one who believes that classification reveals order, a number of desiderata suggest themselves for classificatory distinctions. First, distinctions should, as far as possible, be sharp and practically decisive. That is to say, it should be possible to determine on which side of a distinction a thing falls, and preferably it should be possible in practice, not merely in principle. In the most orderly universe possible, the ultimate distinctions would be completely sharp.

Second, to play any role in supporting a metaphysics of order, correct classifications must, in some sense, be discovered rather than merely invented. This criterion is independent of the previous one. For example, the comprehensive classification of things in the world into those weighing under 1 kg, those weighing 1–2 kg, and so on, would be both sharp and readily applicable. But intuitively it would be thoroughly artificial: we would surely not imagine that such a classification contributed in any way to our understanding of any preexistent features of the things in the world. On the other hand this intuition is not altogether easy to clarify. Certainly the proposed classification is real in the sense that it depends on real, objective properties of the objects.

What appears to be missing is not so much reality—or even existence independent of our classifying activity, since presumably things do have weights whether or not we know it—but significance. Knowing that an object weighs between 5 and 6 kg tells us virtually nothing else about it. And this brings us to perhaps the most important desideratum of a classificatory scheme, that assigning an object to a particular classification tell us as much as possible about that object. For an extreme conception of order, the ideal would be that correct classification should potentially tell us everything about an object. Of course, however similar two objects might be, their behavior might differ as a result of differences in circumstances. So "everything" must be restricted to something like intrinsic properties and dispositions. Even this is perhaps a more ambitious goal than anyone would seriously hold. At any rate, rather than chip away at this ideal of perfectly orderly kinds, I propose to defend a diametrically opposed picture. My thesis is that there are countless legitimate, objectively grounded ways of classifying objects in the world. And these may often cross-classify one another in indefinitely complex ways. Thus while I do not deny that there are, in a sense, natural kinds, I wish to fit them into a metaphysics of radical ontological pluralism, what I have referred to as "promiscuous realism." This account of natural kinds will provide the starting point for my general thesis of metaphysical disorder.

Science and Ordinary Language

Classification is not merely a part of the professional practice of science. On the contrary, classification pervades any descriptive use of language whatever. A meaningful predicate must distinguish, more or less precisely, what falls under it from what does not, thereby, in a sense, classifying the things in the world. Closer to what we naturally think of as scientific classification, and to my present concern with questions about the *kinds* to which things belong, many people quite innocent of scientific training have mastered quite complex schemes of classification of naturally occurring objects. This fact raises a theme that will be important at several points later in this book, the relation between scientific

understanding and that provided by—perhaps quite sophisticated—common sense.

The ontology of common sense is highly pluralistic. Common
sense imposes order on the buzzing, blooming confusion of phenomena not by unifying them under a relatively simple structure
of fundamental concepts, but by piecemeal extension of knowledge. A model of what I am thinking of as an instrument of
sophisticated common sense here would be a good television nature documentary. From such documentaries much can be learned:
the appearance and gross characteristics of kinds of organisms to
the point where they could readily be identified; aspects of their
behavior, testified to—subject to more or less interpretation—by
photographic evidence; and, through the inevitable and pervasive
hypothesizing of evolutionary or at least adaptive explanations, a
general idea of the broad strategies with which contemporary
science approaches certain kinds of biological issues. Despite this
scientific gloss, such acquisition of information does little, if anything, to unify our knowledge. Rather it seems just to add more
kinds of things and more truths to our antecedent awareness of
the diversity of nature. Thus one who thinks *scientific* classification
should reveal order and unity in the world will presumably want
to distinguish sharply between scientific categories and the prescientific categories of everyday life.

To pose the issue more concretely, a television documentary
on the three-toed sloth should amply equip the attentive viewer
to distinguish perfectly any three-toed sloth from any other object
she or he might encounter. We might then ask whether such a
viewer has thereby learned a modest piece of science, expanded
his or her extrascientific knowledge of the world, or, finally,
whether this is not a thoroughly bogus dichotomy, grossly ex-
aggerating the distance between scientific and nonscientific ways
of understanding. Since biological organisms provide a domain in
which both scientists and nonscientists frequently acquire quite
elaborate classificatory abilities, it is an ideally suited domain for
examining the relations between scientific and nonscientific categories.

Before this topic is pursued further an important philosophical
debate must be mentioned. It has recently been held that, from a

biological point of view, terms apparently denoting kinds of biological organisms do not denote kinds at all, but rather individuals.[1] Specifically, it is claimed that such terms should aim to denote chunks of the genealogical nexus, united by common relations of descent. If this thesis is accepted, then it appears that the seeker after categorial order must ascend to some higher level of abstraction, for example to some category of which species, genera, and so on are merely instances. Thus it is sometimes supposed that species are a kind of entity subject to some kind of natural law *qua* species. It might also be imagined that genera, families, and so on provided a hierarchy of such kinds of entities, although this seems improbable in view of the lack of any consensus as to how these higher-level entities should be distinguished. Or, finally, there might be a kind consisting of all chunks of the genealogical nexus of whatever size. It will be apparent that such a higher-level category must abstract from some vast differences between its members if it is to incorporate not only the diverse biological species but, presumably, the higher-order groupings (taxa) as well, for example Aves, the object consisting of all and only the birds living, dead, or yet to be born.

The merits of the biological kinds as individuals thesis will be considered in more detail in the next chapter. For now, I shall make just two points in defense of the present treatment of biological kind terms as just what they appear to be. First, the thesis that species are individuals is grounded entirely in considerations of evolutionary theory. Other parts of biology seem much less hospitable to the suggestion. For example, ecology, involving the interactions between different *kinds* of organisms, seems to be impossible to formulate in any but the most abstract terms without treating specific kinds of organisms as kinds. (The cynic might reply that that is why so much of this field is indeed so abstract as not to seem likely to apply to anything so vulgar as a real group of organisms.) And second, my present goal is to consider the relations between the use of classificatory terms in biology and everyday language. No one can seriously doubt that, as terms of ordinary language, *cat, cow, dandelion,* and so on are perfectly normal general terms. If they wholly change this character when taken rather as scientific terms the comparison at which I am aiming will be impossible to make. This seems intelligible only if there is a truly unbridgeable gulf between science and the rest of human knowledge. If, as I want to claim, there is no such gulf, then at least some account will be required for the anomalous case of biological kinds. It seems reasonable to begin the discussion, at least, by assuming that terms for biological kinds are roughly homonymic between scientific and extrascientific uses.

An obvious way of resolving the problem of relating scientific to nonscientific categories is to suggest that the latter are crude approximations to the former, and indeed may be guided toward convergence with the former as scientific sophistication increases. So, it is often suggested, a term such as *fish* was gradually weaned by scientific insight from its erroneous inclusion of whales. A strategy along these general lines has been widely adopted in the last twenty years as a result of the influence of Hilary Putnam and Saul Kripke.[2] The idea has been developed in most detail by Putnam, and I shall here focus on his exposition of it.

To introduce Putnam's theory we need first to recall a distinction made famous by Locke, that between real and nominal essences. Whereas the former is, roughly, whatever accounts for the characteristic nature of things of a certain kind ("the being of anything whereby it is what it is"), the latter is merely the feature or set of features that we use to distinguish objects as belonging to the kind ("the abstract idea which the general, or sortal . . . name stands for").[3] For something such as a triangle, for Locke a wholly conceptual object, the real and nominal essences coincide. Since the properties of a triangle flow solely from the way it is defined, contemplation of the latter could provide insight into the former. But the distinction was designed, in part, to show the futility of the scholastic, contemplative view of science. Contemplation of forms, nominal essences if anything, would be a source of knowledge of real things only if nominal essences were also real essences. But they are not, so it is not. In the case of material things Locke, like his successors, thought that the real essence must be some feature of the microscopic structure—that the microscopic structure was the real source of the phenomenal properties of a thing, and that microstructural properties accounted for the homogeneity of macroscopic kinds. Of the practical value of this notion, on the other hand, Locke was skeptical. Regretting, famously, our lack of microscopical eyes,[4] he doubted whether real essences, if they were, *per impossibile,* discovered, would coincide with the merely nominal kinds we had so far discovered. Thus he held that kinds of things were demarcated by nominal essences only.[5] Subsequent scientific history has convinced some, perhaps most, philosophers that Locke's skepticism was premature. Chemistry and physics have revealed a good deal about the microstructure of things, and the members of many antecedently distinguished classes of things have been found to share important structural properties.

This optimism is well represented by the ideas of Putnam and Kripke. Putnam maintains that genuine natural kinds provide the extensions of many terms of natural language, where these natural kinds are determined by true Lockean real essences. Putnam's theory of the meaning of such terms incorporates, and sharply distinguishes, equivalents of both real and nominal essences. But it is the real essence that determines the extension of the term; and it is this role that offers the reconciliation between scientific and natural language terms which is my present concern, but which I believe cannot, in the end, be successfully sustained.

In this chapter I shall use the term *natural kind* to refer to a class of objects defined by common possession of some theoretically important property (generally, but not necessarily, microstructural).[6] The traditional view, to which Locke may be counted a subscriber, is that terms of ordinary language refer to kinds whose extension is determined by a nominal essence, and hence not, in the present sense, to natural kinds.[7] Optimists, at least, believe also that science, by contrast, attempts to discover those kinds that are demarcated by real essences. It is compatible with these two views that on occasion real and nominal essences might coincide. But this would be largely fortuitous. For a number of reasons such a position does not exclude important connections between science and everyday language. First, the explanation of our recognition of a kind might trace back to a theoretical feature that defined a true natural kind. Second, terms that originate in purely scientific discourse often find their way into ordinary language. And third, it is widely accepted that even the most straightforwardly observational terms are to some degree "theory-laden," although the exact extent of this is much debated.[8] But the general picture is of science as a largely autonomous activity in spite of subtle and pervasive interactions with the main body of language. One of the attractions of Putnam's essentialism is that it promises to provide much stronger links between science and ordinary langauge, since many terms of the latter are claimed to refer to kinds demarcated by the former.

Putnam's theory resolves the meaning of a natural kind term into four components, referred to as a syntactic marker, a semantic marker, a stereotype, and an extension.[9] For example, the term *elephant* might have as syntactic marker "noun," as semantic marker "animal," as stereotype "large gray animal with flapping ears, a long nose, and so on," and an extension determined by the microstructural (or other theoretical) truth about elephants. It is with the last two of these, which are approximately equivalent to nominal and real essences (the stereotype being the nominal essence, stripped of its reference-fixing function), that I shall be concerned.

The distinction between the stereotype and the extension is reflected in a distinction between mere competence in the use of a term, and full knowledge of the meaning of a term. The former requires only the first three components of meaning. In fact, the stereotype is explained as being the set of features that must be known by any competent speaker of the language, regardless of whether it provides a good guide to the actual extension of the term.[10] All this ignorant talk is facilitated by what Putnam calls "the division of linguistic labor."[11] If, for any reason, it is important that items be assigned to the correct classes, it is necessary that there be experts, familiar with the really essential properties of the kinds in question, who are thereby able to perform this function. Although some may be taken aback at the idea of appealing to experts to adjudicate the meaning of our unstudied talk, we should recall that in the case, for instance, of gold, we do generally have to take it on authority whether or not some object in our experience is the genuine article. We should note, however, that we can never be sure whether even the experts fully know the meaning of the term. For there is no guarantee that they have yet got right the real essence of the kind in question. (Devotees of the so-called pessimistic induction on the history of science might find it particularly distressing to discover that the provisionality of science here threatens to undermine the intelligibility of our everyday speech.)

The central question raised by Putnam's analysis is how the nominal, or stereotypic, kinds of ordinary language are to be correlated with the natural kinds discovered by science. Or conversely, granted that there are these real, empirically discoverable natural kinds, how do we know which to assign to a particular term? Putnam answers this question by appealing to a previously unnoticed indexical component of meaning.[12] This consists in the reference, in using a natural kind term, to whatever natural kind paradigmatic instances of the extension of the term "in our world" belong to. Such a paradigm may be identified either ostensively, or operationally through the stereotype. Once the paradigmatic exemplar has been identified, the kind is then defined as consisting of all those individuals that bear an appropriate "sameness relation" to this individual. The sameness relation is Putnam's analogue of Locke's real essence.

It should be clear that Putnam's theory offers precisely a way of tying our ordinary language classifications to those provided, or eventually to be provided, by science. Given Putnam's picture it is natural to suppose that as science advances, ordinary language will be adjusted (in the manner of the dewhaling of fish) to conform to these more accurate categories. Categories that turn out to be, from a scientific point of view, wholly wrongheaded may be abandoned. Or if they are not, we can perhaps recognize that certain backwaters of ordinary language carry on outside the pale of orderliness described by science (perhaps as a resource for poets and the like).

Putnam's theory, however, is wholly untenable. While I readily agree that knowing the meaning of a general term is something that admits of degrees, and that the higher degrees may sometimes be accessible only to experts, I do not think that experts can deliver on quite the task Putnam sets for them. And the task on which they can deliver is different in degree rather than kind from what can be expected from a linguistically competent nonexpert.[13] Most fundamentally, as a theory of biological kinds, Putnam's theory founders on the complete absence of the sameness relations its application would require. Before substantiating these claims, I shall say a little about the kinds of arguments offered by Putnam (and to some extent Kripke) in support of this kind of theory.[14]

Putnam's general methodology is to consider counterfactual situations in which we encounter an item that is in some interesting respect novel, and then to decide (intuit?) whether we would apply a particular term to it. The relevant cases are of two kinds: first, those in which an object satisfies the stereotype but for theoretical reasons is excluded from the extension of the term; and second, those in which all the theoretically important conditions are met, but part or all of the stereotype is not. I shall concentrate on cases of the first sort.

A favorite example of Putnam's is set in a place called "Twin Earth." This now well-known location is identical with Earth in every respect, except that what is there called "water," a substance that plays exactly the role that water does on Earth and shares all the phenomenal properties of Earth water, turns out not to have the chemical composition H_2O, but to be some other complex chemical substance, which may be called XYZ. Putnam's contention is that when we discovered this fact we would have to say that what they called "water" was not water, since water is, necessarily, H_2O. Being H_2O is what constitutes the sameness relation for the natural kind water. Since we have discovered that this is the appropriate sameness relation in our world, this fact has been incorporated in the very meaning of the term. The first point I wish to emphasize here is a methodological one. If Putnam says "XYZ is not water," and my intuition is that it would be (another kind of) water, how is such a dispute to be settled? Who knows what we ought to say in such a fantastic situation? Of course, the claim that XYZ would not be water must itself be intuitively compelling if it is to support, not merely illustrate, Putnam's theory.[15]

Perhaps it will be helpful to notice that scientific history encompasses similar, if less extreme, cases. Consider, for instance, the first European botanists to study North American trees. On arrival they might have been interested to discover that there were beech trees on that side of the Atlantic. More careful investigation would have taught them that these beech trees were significantly different from those that they had previously encountered, and should be assigned to a different species.[16] Since the most striking difference between the two species was (perhaps) the difference in the size of the leaves, this discovery was commemorated in the distinction between *Fagus sylvatica* and the newly discovered *Fagus grandifolia*. In view of overwhelming similarities, there could have been little doubt about assigning these trees to the genus *Fagus* (= beech). Let us suppose that one of our botanists was also a linguist. If a native had asked her whether there were beech trees where she came from, what ought she to have said? My intuition, for whatever it is worth, is that she should have said that there were; though naturally if she were talking to a native *botanist,* she would go on to add that European beech trees belonged to a different species.

The purpose of this example is to suggest that plausible though some of Putnam's examples may be, they do admit of different interpretations. In the case of the water example it is also important to emphasize the great improbability of Putnam's hypothesis, and the consequent indeterminacy as to how we would accommodate such a circumstance. All our scientific experience goes against the possibility of there being two substances differing solely in having radically different molecular structures.[17] But this should not blind us to the fact that if we do take the possibility seriously, the best way of accommodating it might be to admit that there are natural kinds that encompass such radical differences of structure. After all, it is surely the absence of experiences like the one Putnam describes that makes it reasonable to attach to molecular structure at least most of the importance that Putnam ascribes to it.[18] Perhaps no one will be persuaded to take the case in the way I have suggested. But I hope that I have at least said enough to motivate a closer look at how such issues are, and can be, treated in both scientific and everyday practice.

Biological Kinds

Putnam's theory requires that there be kinds discriminated by science that can provide the extensions of certain terms of ordinary language. In consideration of this proposal, I shall look at classificatory terms in biology. The part of biology that is concerned with the classification of organisms is taxonomy. Within taxonomy an organism is classified by assigning it to a hierarchical series of taxa, the narrowest of which is the species[19]. Thus a complete taxonomic theory could be displayed as a tree, the smallest branches of which would represent species. Rules would be required for assigning individual organisms to species, and an individual assigned to a particular species would also belong to all higher taxa in a direct line from that species to the trunk of the tree. Let us assume what might be called "taxonomic realism." This is the view that there is one unambiguously correct taxonomic theory. At each taxonomic level there will be clear-cut and universally applicable criteria—essential properties, let us say—that generate an exhaustive partition of individuals into taxa. We can also assume that the appropriate number of taxonomic levels to recognize is somehow implicit in the nature of organisms. The claim that there are natural kinds in biology, demarcated by real essences (and *a fortiori* Putnamian sameness relations), would thus be wholly sustained. Even under these circumstances, as I shall now show, Putnam's theory faces severe difficulties of application.

The central difficulty is that it is far from universally the case that the preanalytic extension of a term of ordinary language corresponds to *any* recognized biological taxon. Taxonomic practice may, of course, change. But, as some of my examples will show, such divergence between scientific terms and ordinary language terms often occurs for good reasons that preclude any reasonable expectation of eventual convergence between the two. In a sense this divergence is not easy to substantiate fully, because the terms in question are often quite vague, and their application somewhat indeterminate. However, I think even this indeterminacy can be seen to corroborate my thesis.

The richest source of illustrations is the vegetable kingdom, where specific differences tend to be much less clear than among animals, and considerable developmental plasticity is the rule. Any observant person who has explored the deserts of the American Southwest will have little difficulty distinguishing a prickly pear from a cholla (the distinction being based on the flattened leaf segments of the prickly pear). Yet taxonomically all the species of both these kinds of cacti belong to the same genus, *Opuntia*. Several species of this genus are surely, to the casual observer, prickly pears, and several are certainly chollas. Taxonomy does not identify any important relation between *Opuntia polyacantha* and *Opuntia fragilis* (two species of prickly pear) that either does not share with *Opuntia bigelovia* (a species of cholla). Ordinary language does make such a distinction, and on the basis of perfectly intelligible and readily perceptible criteria. Thus the property of being a prickly pear is just not recognized in biology.

Or consider the lilies. Species which are commonly referred to as lilies occur in numerous genera of the lily family (Liliaceae). To take a few examples again from the flora of the Western United States, the lonely lily belongs to the genus *Eremocrinum*, the avalanche lily to the genus *Erythronium*, the adobe lily to the genus *Fritillaria*, and the desert lily to the genus *Hesperocallis*. The white and yellow globe lilies and the sego lily belong to the genus *Calochortus*; but this genus is shared with various species of mariposa tulips and the elegant cat's ears (or star tulip). I would not want to undertake the task of describing the taxonomic extension of the English term *lily*. However, it is fairly clearly well short of including the entire family. To include the onions and garlics (genus *Allium,* and, incidentally, another excellent example of the point of the previous paragraph) would surely amount to a debasement of the English term.[20]

It is not hard to find similar examples in the animal kingdom. The various species of chickadees and titmice share the same genus. Hawks probably make up three of the four families in the order Falconiformae, although there are some questionable subfamilies. Whether a kite, an eagle, or a caracara is a hawk is a futile debate that I shall not attempt to initiate; but I feel sure that a vulture is not. A particularly interesting example is provided by the moths. The order Lepidoptera includes the suborders Jugatae and Frenatae. It appears that all the Jugatae are moths. The Frenatae, on the other hand, are further subdivided into the Macrolepidoptera and the Microlepidoptera. The latter seem again to be all moths. But the former include not only some moths but (all) skippers and butterflies.[21] The trouble here is not that we cannot give a reasonably plausible account of the extension of the English term *moth,* but rather that the grouping so derived is, from a biological point of view, quite meaningless.

One other example from the animal kingdom that will be taken up in a later chapter is the case of rabbits and hares. Taxonomically these have the same relation as chollas and prickly pears, all belonging to species of the same genus, *Lepus.* Physiologically, the animals are extremely similar. Nevertheless, the distinction between them is clear enough to the farmer or hunter. As Thomas Bewick nicely described it:

> Notwithstanding the great similarity between the Hare and the Rabbit, Nature has placed an inseparable bar between them, in not allowing them to intermix, to which they mutually discover the most extreme aversion. Besides this, there is a wide difference in their habits and propensities: the Rabbit lives in holes in the earth, where it brings forth its young, and retires from the approach of danger; whilst the Hare prefers the open fields, and trusts to its speed for safety.   (Bewick, 1824, p. 374)

A rather desperate attempt might be made to save the theory from such examples by going for the best available taxon and accepting some revisionary consequences for ordinary language. Thus one might claim that the extension of "lily" was the whole family Liliaceae, or of moths the order Lepidoptera. We would just have to accept the fact that onions had turned out to be lilies, or butterflies moths. In defense of such claims, it could be pointed out that ordinary language has indeed come to accept scientifically motivated changes. Again, the stock example is the exclusion of whales from the extension of the term *fish* consequent on the discovery that they are mammals. But actually this example is by no means as clear-cut as is sometimes assumed. In the first place, *mammal* is more a term of relatively sophisticated biological theory than of prescientific usage. Fish, by contrast, is certainly a prescientific category. What is more doubtful is whether it is genuinely a postscientific category, for it is another term that lacks a tidy taxonomic correlate. I assume that the three chordate classes Chrondichthyes, Osteichthyes, and Agnatha would all equally be referred to as fish (unless sharks and lampreys are just as good nonfish as whales). But unless there is some deep scientific reason for lumping these classes together and excluding the class Mammalia, the claim that whales are not fish is a debatable one. Perhaps *fish* just means aquatic vertebrate, so that whales are both fish and mammals, and this well-worn example is just wrong. However, whales were never the most stereotypical fish, and it is easy to see the point of denying that they are fish at all: they do belong to a taxonomically respectable group most members of which do not remotely resemble fish. Suggestively, it appears that recently the distinction has become a highly significant one on moral grounds. Consider the importance attached to catching tuna without killing dolphins. At any rate, I see no parallel considerations in favor of the claim that butterflies are moths.

The second difficulty for the application of Putnam's theory that occurs even against the background of taxonomic realism, concerns the hierarchical structure of taxonomic theory. Putnam's theory determines the extension of a natural kind term by means of a theoretical sameness relation to a suitable exemplar. Suppose we want to discover the extension of the word *beetle*. A suitable exemplar would have to satisfy the condition that it be readily recognizable as a beetle by a linguistically competent layperson; but probably that would not eliminate a very large proportion of the approximately 290,000 recognized species.[22] Any particular exemplar will belong (or at least be assigned) to one species. Given taxonomic realism, there will then be some sameness relation that it displays to other members of that species, some relation that applies within its particular genus, and so on up, not just to the relation that holds between all members of the order Coleoptera, which is approximately coextensive with the term *beetle,* but beyond, as far as the relation that holds between it and all animals but no plants. One may well wonder how the appropriate sameness relation is to be selected from these numerous alternatives.

A kind of solution to this difficulty does suggest itself. If we collected a sufficiently large number of beetles, as different from one another as was consistent with the stereotype, we could look for the narrowest sameness relation that held between every pair of our specimens. This methodology would, of course, force us to identify moths with the order Lepidoptera and to accept the consequence that butterflies were a kind of moth. It also seems to me that collecting the set of samples would involve attaching a lot of significance to the stereotype; if the stereotype were not a good guide to the real extension, it could hardly work. Rather than pursue this somewhat farfetched suggestion, however, it is time to treat the assumption of taxonomic realism more critically.

Despite the preceding examples, I do not mean to deny that many general terms for living organisms do have a reasonably clear taxonomic correlate. For almost all species of birds and large vertebrates, for many flowering plants, and for some species of fish and insects, there is something (or sometimes a list of things) referred to as a common name. It is not obvious whether these should be thought of as part of ordinary language or as part of a technical vocabulary. Certainly if competence in English does not require enough biological expertise to distinguish a beech from an elm,[23] then surely it cannot require an awareness even of the existence of the solitary pussytoes, the flammulated owl, or the Chinese matrimony vine. (Given that such terms are in fact as much or more the special province of the birdwatcher or wild-flower enthusiast, whose activities are generally treated with con-siderable disdain by professional biologists, as of the biologist, they also serve nicely to threaten any sharp dichotomy between the language of science and the language of everyday life.) If such charming terms are assigned with their Latin equivalents to sci-entific taxonomy, and if we restrict our attention to terms with which the layperson can reasonably be expected to be familiar, then one thing we will discover is that where there is a recogniz-able corresponding taxon, it is generally of higher level than the species.

For the case of large mammals, where human interest and empathy are at their highest, most familiar terms do refer to quite small groups of species; and common specific names are often widely known (such as blue whale, Indian elephant, or white-tailed deer). Most well-known names of trees refer quite neatly to genera, as, for example, oak, beech, elm, willow. An interesting exception is provided by the cedars. The various species referred to by this name are not closely related. It is natural to suppose that the term *cedar* has more to do with a kind of timber than with a biological kind. With birds, the situation is highly varied. Ducks, wrens, and woodpeckers form families. Gulls and terns form subfamilies. Kingbirds and cuckoos correspond to genera, while owls and pigeons make up whole orders. The American robin, finally, is a true species, although, interestingly, in Britain *robin* refers to a quite different species, and in Australia, I have been told, it refers to a genus of flycatchers. For insects, where the number of species is much greater, and the degree of human interest generally lower, the mapping is predictably coarser. Such things as hump-backed flies, pleasing fungus beetles, brush-footed butterflies, and darkling beetles make up whole families (the last-named, for instance, having some 1,400 known North American species). Here we are already close to a technical scientific vocabulary. More familiar things such as beetles and bugs encompass whole orders. (For that matter, must the competent speaker of [American] English know that a beetle is not a bug? Or is the word *bug* ambiguous?)

One reason why it is significant that if ordinary language terms correspond to any taxon it is generally a taxon of higher level than the species is that whereas the ontological status and reality of species remain much-debated issues, there is little reason to believe in anything like taxonomic realism for any higher level of classification. Although there are heated debates about the appropriate arrangement of species into genera, families, and so on, these debates seem almost entirely pragmatic rather than deeply theoretical. Thus, insofar as kind terms in ordinary language correspond, more or less, to these higher-level taxa, there appear to be no more "scientific" kinds that could possibly provide the sameness relations, or essences, required by Putnam's theory.

A qualification must be entered at this point. The ultimately arbitrary nature of higher-level classifications is at its least controversial within the taxonomic tradition grounded ultimately in morphological features of organisms. However, in recent years such a conception of taxonomy has been increasingly superseded by a phylogenetic conception of classification, one in which classification is supposed to reflect the facts of evolutionary history.[24] Although there are major differences within this general conception, one very influential school, cladism, insists that every taxonomic distinction should reflect an evolutionary event of lineage bifurcation, and thus that every taxon should include all and only the descendants of one side of such an evolutionary division. Pending a more detailed discussion of these issues in the next chapter, the following point is sufficient for present purposes. It is true that from a cladistic point of view groups at all different levels will be equally real. Indeed every taxon has the same claim to existence as every other. However, terms of ordinary language have a small chance of referring to cladistically acceptable taxa. The problem can be illustrated with a stock example for showing the difference between cladistically acceptable and unacceptable taxa. For a cladist, neither the reptiles of ordinary language nor the Reptilia of the traditional systematist can be acceptable taxa. This is because neither reptiles nor Reptilia include birds (Aves). But, or so we believe, birds evolved from some primitive reptile. But then any proposed ancestral group that could legitimate the taxon Reptilia would necessarily include, as descendants of the very same ancestral group, Aves.

I think it is clear that this strong phylogenetic condition is one we have no reason to expect that our ordinary language terms can meet. This is simply because such terms surely originate in morphological characteristics rather than in detailed phylogenetic knowledge. *Reptile* is a case in point. Nothing in the preceding discussion even begins to suggest any defect in this term, at least for the ordinary purposes of ordinary language, nor any reason why we should wish to start referring to birds as reptiles. The discussion above of cases such as lilies provides instances of the same problem. But with cladistic taxonomy we move from such morphologically motivated divergences from scientific taxonomy—for example, the importance we attach to the particular taste of certain species of Liliaceae—to divergences that may be quite invisible. From a cladistic point of view it may well turn out that, in accordance with the latest evolutionary doctrine, a lily that would be perfectly appropriate for a funeral bouquet turns out not to be a lily at all. There may be a case for such linguistic revisions in professional systematics, but there can be no possible case for holding ordinary language hostage to such quirks of evolutionary history. Thus, in summary, we may choose between a theory of taxonomy that attaches no theoretical importance to higher taxa, and one in which the distinctions among higher taxa are motivated by considerations quite unrelated to any of concern to ordinary language classifications. In either case, to the extent that the candidates for the "scientific" correlates of ordinary language terms are taxa above the level of species, a Putnamian essentialism will be wholly inapplicable.

The Functions of Classification

A natural way to contrast the classificatory schemes of ordinary language and of scientific taxonomy is in terms of the very different functions they serve. Ordinary language has a variety of aims in distinguishing kinds of organisms, and these need not coincide on the same classifications. In the next chapter I shall argue, in extreme contrast to taxonomic realism, that the same is true even for scientific taxonomy.

Ordinary language classifications are, unsurprisingly, overwhelmingly anthropocentric. A group of organisms may earn recognition in ordinary language for any of various reasons: because it is economically or sociologically important (Mediterranean fruit flies, silkworms, or tsetse flies); because its members are intellectually intriguing (trap-door spiders or porpoises), furry and appealing (hamsters and Koala bears), or just very noticeable (tigers and giant redwoods). The fact that this list could be extended almost indefinitely merely reflects the immense variety of human interests and the number of these that involve, to some degree, other living beings. From this perspective, many apparent anomalies between ordinary language and scientific terms are readily explicable. To recall an earlier example, it would be a severe culinary misfortune if no distinction were drawn between garlic and onions, a distinction that is not reflected in scientific taxonomy. Presumably there is no reason why taxonomy should pay special attention to the gastronomic properties of its subject matter.

Consider a slightly more elaborate example. The taxonomic classes birds and mammals are both parts of ordinary language (though the latter less clearly). By contrast, the much more numerous class of angiosperms (flowering plants) receives no such recognition. There is a very familiar term of ordinary language, *tree*, the extension of which undoubtedly includes oak trees and pine trees, though perhaps not their seedlings or even saplings. The extension of the term *angiosperm*, on the other hand, includes daisies, cacti, and oak trees but excludes pine trees. It is no surprise that such a grouping finds few uses outside biology: for most purposes it is much more relevant whether something is or is not a tree than whether its seeds develop in an ovary. This seems amply to explain why the term *angiosperm* has no place in ordinary language. Similarly, the gross morphology of a plant is of little importance to taxonomic theory; hence the term *tree* has no place in scientific taxonomy (although it might be important to aspects of ecology).

Organisms that are of little interest to nonspecialists are typically coarsely discriminated in ordinary language. Thus it is that despite the vastly greater number of arthropod than vertebrate species, ordinary language discriminates many more kinds of the latter. Vertebrates are more likely to be useful (nutritious, furry), appealing (cute, furry, intelligent) or just noticeable (big, frightening). Lacking these features, arthropod classifications in ordinary language tend to cover enormous numbers of species. "Beetle" is probably as fine a discrimination as the majority of competent speakers of English can make, and even this, if correctly distinguished from other common arthropods, shows a degree of biological sophistication. Pursuing this broadly functionalist perspective in a slightly different direction, we should anticipate the likelihood that there are other, specialized vocabularies that coincide neither with ordinary language nor with the vocabulary of the systematist. The vocabularies of the timber merchant, the furrier, or the herbalist may involve subtle distinctions among kinds of organisms; there is no obligation that these coincide with those of the taxonomist. The term *cedar* is very likely a case in point.

Scientific taxonomy, it is to be hoped, avoids to some extent this anthropocentric viewpoint. The number of species named is intended to reflect the number of species that exist. Nonetheless, even here there may well be unavoidable anthropocentrism. There is considerable and legitimate uncertainty about what constitutes the distinct existence of similar but related species. To the extent that this is ineliminable, it seems likely that the distinctions that are drawn will, as in the case of ordinary language distinctions, depend to some degree on the purposes for which they are drawn. And if these purposes are various, so may be the useful schemes of classification. Moreover, there is no conception of species that supports the idea that there is any privileged sameness relation demarcating the members of a species. Thus the criteria chosen in practice to distinguish the members of a species are likely to be chosen in part for anthropocentric reasons such as ease of human application. And this may on occasion affect details of the taxonomic system.

I shall elaborate and defend these claims in the next chapter. For now, I shall advocate as a general reaction to the phenomena I have been considering the position that I have referred to as promiscuous realism. Nothing I have said, either about scientific kinds or about the kinds of ordinary language, suggests that these kinds are in any sense illusory or unreal. They may typically have vague boundaries, to be sure, but that is a quite different matter. Hence the realism. The fact that the kinds of ordinary language do not coincide with those of scientific taxonomy is enough to constitute a modest pluralism, though perhaps to characterize it as "promiscuous" would seem an exaggeration. The beginnings of the promiscuity can be seen in my claim that ordinary language has a variety of grounds for classification that should be expected further to cross-classify the biological world. Many who might be quite willing to concede this point will perhaps find it of little fundamental importance so long as science, at least, can be counted on to provide an orderly refuge from this extrascientific Babel. However, in the next chapter I shall argue that there is no such refuge, and that pluralism is as plausible within science as it is between science and the rest of language. That argument will complete my case for a truly promiscuous realism.

2. Species

In the last chapter I argued that very few terms of ordinary language could be taken to refer to biological species, or in many cases to any other group recognized by biological science. Although this conclusion leaves a pressing question about the relations between the classificatory terms of science and ordinary language, it also leaves it open to the believer in the orderly world disclosed by science to respond: So much the worse for ordinary language. Perhaps all we are discovering is the fundamental confusion endemic to prescientific discourse. This diagnosis would fit well with a prominent strand of contemporary scientistic philosophy, most conspicuously exemplified by the eliminative materialism of Paul and Patricia Churchland, who argue for the fundamental confusion of most or all of our common talk about the mental.[1] However, to sustain this response, it is necessary that when we look at the scientific understanding of species, and preferably even the scientific construction of the further edifice of higher taxa, we should come across a determinate and orderly realm of phenomena. What I shall argue here, on the contrary, is that the situation when we attempt to assign organisms to species is quite as messy as that revealed for the case of nonscientific classification. So that rather than science providing an orderly resting place for the preliminary confusions of ordinary language, the complexity of ordinary language provides a model for the more technical classificatory exercises of scientists.

Species are, by definition, the lowest-level classificatory unit, or basal taxonomic unit, for biological organisms. Although further subdivisions into subspecies, races, or varieties are sometimes used, these are generally assumed to be of little theoretical importance. It is also often supposed that species are a uniquely significant level of division of biological kinds; although classification extends to much broader groupings (genera, families, and so on), these are often taken to be arbitrarily chosen in a way that species are not.[2]

In this chapter I shall look at three issues concerning the nature and status of species. First, there is a major disagreement about the very ontological category to which species belong: are they kinds, sets, individuals, or what? Second, there are a number of very different accounts extant of what constitutes membership of a species; most fundamentally, these debates concern whether the members of species are connected by evolutionary relations or by the common possession of morphological or physiological features. Third, a very traditional issue is whether there is some essential property defining membership of a species. This issue warrants treatment here because it is currently rather uncontroversial that the acceptance of Darwinism forces the rejection of this aspect of essentialism. On the other hand, as the views of Putnam and Kripke indicate, essentialism in general is by no means dead. Among more biologically sophisticated philosophers essences are not sought at the level of species, but are thought of as defining more theoretically refined categories. The discussion of the failure of essentialism at the level of species will, therefore, set the stage for a more general discussion of essentialism in the following chapter. Finally, the question of whether species have essences has some tendency to be conflated with the question whether species are real. Insisting on the distinctness of these questions will help to clarify further what I mean by promiscuous realism.

The Ontological Status of Species

It should be no surprise that recent thinking about the nature of species has been driven largely by the attempt to reconcile our understanding of the species concept with the theory of evolution. Given, first, that *species* is to be understood as a theoretical term; second, the widely agreed impossibility of understanding theoretical terms in the absence of a proper understanding of the theoretical context in which they occur; third, the dominant role that the theory of evolution plays in so much of biological theory; and fourth, the fact that the evolutionary process is assumed to provide, in a well-known phrase, the origin of species, it would be remarkable if the concept of species were not strongly dependent on the needs of evolutionary theory. This connection with evolutionary theory is the main source of arguments for the rather counterintuitive thesis that species are not kinds at all, but rather are individuals.[3] On this view, to refer to something as, say, a pig is not to say anything directly about what kind of thing it is, but rather to say what larger, more enduring thing it is part of, in this case the species *Sus domesticus*. This latter thing is then to be understood as a particular chunk of the total genealogical nexus of life on Earth. Species membership is thus seen as requiring, among other things, the appropriate spatiotemporal location. However piglike the creatures we might discover on Alpha Centauri, and however readily they might interbreed with terrestrial swine, and although their genetic code might be indistinguishable from that of an Arkansas razorback, they would not and could not be real pigs.

A number of arguments have been offered in support of this thesis. A starting point for the Darwinian theory of evolution is that the members of species vary. On a very restrictive conception of natural kinds, one that required absolute homogeneity of the members of a natural kind, this point alone is sufficient to rule out species as belonging to this ontological category. But such a conception is clearly too restrictive. Even atoms of gold or molecules of water, two of the main contemporary paradigms for natural kinds, are not *absolutely* homogeneous. Atoms of a particular element generally come in different isotopic forms, and even if we claim that it is really isotopes not atoms that are the natural kinds (thus divorcing chemistry from ordinary language), atoms are said to differ with respect to such features as energy levels of electrons. For anything more complex the conception is pretty obviously useless.

A more serious argument begins with the observation that a central thesis of evolutionary theory is that species evolve. But a kind, being an abstract object, cannot do anything, including evolve. Evolution, moreover, involves by definition real changes, which abstract objects are surely unable to exhibit. Although the issues here are complex,[4] it seems to me that this argument is also easy to resist. Again consider one of the paradigm natural kinds, water. Water can be created, say by burning hydrogen, or destroyed, as by electrolysis. Properties of individual bits of water can change, for example by getting hotter or dirtier. And it is even possible that all the water everywhere could get hotter and dirtier. (If it should happen that there is no water anywhere but on Earth, this situation is very probably being realized.) In the latter case, surely, we do not need to say that the *kind* water has changed, merely that all the bits of water that make it up have changed. In the case of a species, then, provided only that we can find some way of defining the kind independently of any narrow set of properties of its members, this problem should be similarly soluble.[5]

A rather different ground for denying that species could be natural kinds is the claim that there are no laws true of their members.[6] Since there are certainly true generalizations about the members of species, the force of this claim must depend on how much is packed into the notions of natural kind and of law. Commonly these concepts are taken to be correlative, so that the natural kinds are just those kinds the members of which satisfy laws of nature. So we should focus on laws. Many philosophers still believe that laws of nature must be exceptionless (for example, Davidson, 1967). Since species are now understood as being variable—this, indeed, being a presupposition of evolution by natural selection—most generalizations about the members of a particular species will have exceptions. Contrary to a major tradition in the philosophy of science, "All swans are white" and probably "All ravens are black" are not even true; and even if they were, we would have no reason to assume that they would continue to be so. One can, it is true, find exceptionless such generalizations, for example, "All humans are mortal" or "All pigs have hearts." But pretty clearly this is a trick, since in such cases the "laws" are consequences of much broader generalizations, and if there are laws involved at all they should be formulated at these higher levels. The former example might even be a consequence of the Second Law of Thermodynamics. Generalizations that are appropriately stated at the species level are either false or could become so without violating anything we believe about the nature of species.

The obvious response to this argument is simply to deny that laws must be exceptionless. This response seems to me absolutely correct. First, the kind of knowledge we have, at least about objects larger than those that are the subject matter of physics and chemistry, is almost invariably not universal in form. In biology this is the case whether we are talking about swans or oak trees, or about more sophisticated-sounding kinds like species (in general), phylogenetic lineages, or parasites. This is even more obviously the case for the social sciences or, to take a quite different area, meteorology. This situation is a large part of the reason why so much current work on laws and causality is concerned with probabilistic laws (see Part III, below). Second, as Nancy Cartwright has argued, even our most paradigmatic physical laws are neither exceptionless nor even true (Cartwright, 1983, essay 3). This is because they tell us only what is true *ceteris paribus* or *ceteris absentibus,* and other things are never really either equal or absent. The existence of exceptions, then, is no obstacle to there being laws, albeit modest low-level laws, about the members of species.[7]

A somewhat different argument offered against species as kinds is that whereas a kind is timeless, in the sense that its members might come into being anywhere in space and time, species are timebound and earthbound. Once the dodo has become extinct, it is claimed, it is impossible that there will ever be any more dodos. Of course, given the complexity of a dodo, it is no doubt improbable to the point of practical impossibility that anything indistinguishable from a historical dodo will ever come into being again. The argument, however, claims not physical impossibility but conceptual impossibility. The problem is that this is surely a conclusion from, not a ground for, the claim that species are individuals. Certainly if to be a dodo is to be a part of a particular spatiotemporal lineage, then when that lineage is gone so are the dodos, actual and possible. However, one who believes that dodos form a modest natural kind will simply deny the conceptual impossibility of a renaissance of dodos.

The final argument I shall consider gets back to the starting point of this discussion, the relation of species to the theory of evolution. As I suggested above, *species* is a theoretical term, and the way to understand such a term is to look at the theory (or theories) in which it is embedded. Species, we may then note, are what the theory of evolution is centrally about. Species evolve, diverge, become extinct, and so on, and evolutionary theory tries to discover the causes and effects of these events. Stephen Jay Gould's insistence on the importance to evolution of mass extinctions and Ernst Mayr's emphasis on the formation of peripheral isolates (the geographic splitting of a species) in speciation seem to require the treatment of species as objects for their proper formulation. This is at its most obvious in the case of continuing debates over the possibility of species selection. But if species are treated as objects in central parts of evolutionary theory, and it is to the role in evolutionary theory that we should look to understand the meaning of the theoretical term *species,* then surely we should conclude that species are objects, which is to say, individuals.

This last line of argument seems to me basically correct, with one important reservation. It is just not the case that the parts of evolutionary theory that deal with the events and changes that species undergo is the sole theoretical context in which species figure. Indeed, there is much more to biology than evolutionary theory. Consider, in particular, ecology. Although ecological theory is generally presented in terms of fairly abstract categories (prey, predator, parasite, saprophyte, and so on), the application of such a model to a concrete ecological situation will depend on a classification of the organisms involved. Such classification will draw on general taxonomy so that, for instance, in a particular system lynxes are the predators and hares are the prey. The explananda of such models will be such things as fluctuations of the numbers in these species. I can see no way of understanding *this* theoretical context other than as requiring the treatment of members of species as members of a kind. Moreover, models at this level are also essential to an understanding of evolution. It is participation in such systems that explains how hares got to run so fast, for example.

David Hull (1989, chap. 4) argues for an exclusive adherence to the genealogical perspective in classification. However, he clearly sees that ecology might provide a conflicting scheme, and indeed seems to endorse the view he attributes to Michael Ghiselin that "the species category is strictly a genealogical unit, while niche is an ecological unit. Mixing the two together is sure to cause problems" (Hull, 1989, p. 122). But surely ecologists will be in trouble if they cannot intelligibly claim that some particular species currently occupies some niche in some ecological system, an apparent mixing up of the concepts of species and niche. To the extent that the occupants of a particular niche[8] do not coincide with the members of a particular genealogical line, a possibility widely acknowledged to occur in some cases, ecologists will have reason to favor a method of classification that is not genealogically grounded. Hull cites Ghiselin and Mayr in support of a view with which he concurs, as agreeing that "the actors in the evolutionary play must be genealogical. Only their roles are defined in terms of ecology" (1989, p. 121). But could one not equally well say: "The actors in the ecological play must be ecological. Only their pedigrees are defined in terms of genealogy"? Why assume that the play must be an evolutionary one? Will that not depend on, for example, what time scale we are interested in? Ecological plays take months or years, while evolutionary ones take eons. The problem is that the actors in the short plays are on occasion different from the actors in the longer play that they jointly compose. This, incidentally, is a vivid instance of the kind of metaphysical disunity that underlies the necessity of a disunified science.

I conclude from these considerations that to the extent that we take theoretical embedding as the correct way to consider the question of the ontological status of species, we are driven to a pluralistic answer: in some contexts species are treated as individuals, in others as kinds. I cannot see anything particularly problematic about this conclusion. Consider a different example. For most purposes I have little doubt that I should think of myself as an individual. However, I am also very nearly coextensive with a population of cells all and only the members of which have a particular genotype. If a genic selectionist perspective on the theory of evolution (of the kind popularized by Richard Dawkins [1976]) makes any sense, then there could be theoretical contexts in which this modest kind (in practice, though not necessarily, coextensive with me) played some role. Exactly analogously, ontological pluralism seems the most reasonable position to take on this aspect of the species question.[9]

Criteria of Species Membership

The question to be addressed in this section is what determines the membership of an individual in a particular species. If one takes species to be kinds or sets, this is a question about membership; if they are taken to be individuals, it is a question about mereological inclusion, or part/whole relations. Although one's view on the individual/kind issue will influence the plausibility of various positions on the present question, it will not be sufficient to determine the answer to that question. While there are a very large number of distinguishable views about the criteria of species membership currently discoverable in the technical literature, these can usefully be grouped into three broad categories: morphological, evolutionary, and pluralistic.

MORPHOLOGICAL

Pre-Darwinian conceptions of species characteristically assumed that membership of a species should be determined by morphological characteristics. (Some of these might also have been considered essential.) The post-Darwinian recognition of the omnipresence of variation has tended to put such views in disrepute. More recently morphologically based taxonomy has been revised in the form often referred to as numerical taxonomy (Sokal and Sneath, 1961; Sneath and Sokal, 1973). This movement aimed to ground taxonomy in objective measurements of overall similarity determined by computer analysis of very large numbers of features. Although aspects of its methodology have been incorporated into taxonomic practice, this approach has not been widely welcomed, in part because of philosophical difficulties with the concept of objective similarity, and in part because of a perceived lack of contact with evolutionary theory.

Despite the rather widespread disfavor of morphological taxonomy, most practical classification depends on the determination of reliable morphological diagnostic criteria. Moreover, in large areas of biology where more theoretically favored approaches seem impracticable or inapplicable—especially microbiology and to a considerable extent botany—morphological conceptions of taxonomy remain more respectable. It is, however, quite widely accepted that some independent criteria for assessing the relative significance of morphological features are required. This seems necessary on philosophical grounds to avoid the difficulties with making sense of absolute similarity, and on biological grounds to assure that properties selected have suitable evolutionary or other theoretical significance.

Once we move away from the association of morphological with numerical taxonomy, it is not difficult to find interpretations of morphological taxonomy that promise to make more contact with biological reality. I have already pointed to one such interpretation in the discussion of ecological versus evolutionary criteria in the preceding section. This discussion suggests that a concern with ecology could motivate a basically morphological taxonomy, with the greatest weight attaching to properties that could be shown to have particular ecological significance. A significant role for morphological criteria will emerge at various points in the following pages.

EVOLUTIONARY

The dominant theme in contemporary systematics is undoubtedly the intent to connect taxonomy as directly as possible with evolution. This tendency has two major subdivisions: biological and phylogenetic.

Biological. Owing in considerable part to the influence of Ernst Mayr, perhaps the most widely discussed conception of species membership in recent years has been the so-called biological species concept (Mayr, 1963). This takes a species to consist of a group of organisms connected to one another by actual or possible reproductive links, and reproductively isolated from other organisms. This conception is most centrally motivated by the thought that reproductive isolation is a necessary condition for two groups to evolve independently, and thus lies at the heart of the explanation of biological diversity. In Mayr's thought it is intimately connected with the broader conception of evolution that conceives of speciation as beginning with geographic separation and ending with the establishment of mechanisms of reproductive isolation sufficient to survive even the breakdown of geographic barriers. A recent development of this basic idea, the mate-recognition species concept, emphasizes the mechanisms by which organisms identify conspecifics as potential mates as the fundamental basis of reproductive isolation (Paterson, 1985).

The major difficulty with the biological species concept is its limited applicability. First, and most obviously, it has no apparent application to asexual organisms, every one of which is isolated from every other.[10] Perhaps a more serious difficulty is that in a great many actual cases, especially, but by no means only, among plants, reproductive isolation is fairly weak. This fact has led some biologists to a somewhat more radical dissatisfaction with the biological species concept, which is reflected in a questioning of the overriding role of isolation in divergent evolution.

Apart from isolation from alien genetic material, the other guiding idea behind the biological species concept is that the flow of genes between members of a species will hold together a species. This last somewhat strange expression is intended to mean either of two quite different things. If species are considered as historical individuals, gene flow is what quite literally holds the parts of the individuals together so as to form the whole individual. On the other hand, if species are taken to be kinds, then the only sense in which gene flow might be said to hold the species together would be by causing the properties peculiar and common to the members of a species.

In presenting some problematic cases for the biological species concept, Philip Kitcher (1989) has emphasized that it is very doubtful whether gene flow is sufficient to serve this cohesive function whichever way it is interpreted. Mayr himself, as Kitcher notes, has appealed to epigenetic and homeostatic mechanisms that maintain the genetic unity of the species in the face of the insufficiency of gene flow to serve this end. But such an appeal immediately raises the question whether these epigenetic mechanisms should not be taken as the decisive criteria of species membership. As Kitcher points out, if it is these mechanisms that are in fact the explanation of the unity of the species (though only in the second of the two senses just distinguished), even an additional requirement that the members of the species be connected by historical links would be no more than an ad hoc attempt to insist on the importance of reproductive connection.

It is important to note that in the case just envisaged it would be quite possible for the two criteria considered to contradict each other. There might well be a group of related kinds of organisms distinguished by a number of different epigenetic mechanisms, with the genetic basis of such differences sufficiently similar to allow relatively frequent mutations from one to another. In that case it could happen that the offspring of an organism of one kind could be of a different kind. It would certainly be possible to insist that this must be described as a case of a highly polytypic species. But if the various kinds differed sufficiently in morphology or behavior to make the distinctions between them important for significant biological purposes, this defense of the biological species concept would be little more than a verbal maneuver. The different kinds could be exposed to quite different ecological forces and even, ultimately, undergo quite different evolutionary developments. The crucial point, to which I shall return when I say more about pluralism, is that there is no a priori reason why all the divisions of organisms into kinds should have the same explanation. If different species have different origins, then we should be open to applying different criteria for distinguishing them.

Phylogenetic. Phylogenetic taxonomy aims at a more direct connection with the historical component of evolutionary theory, by starting from the principle that taxonomy should accurately reflect genealogy. Thus a necessary condition for a group of organisms to constitute a species is that they should share descent from some common set of ancestors. The condition is evidently not sufficient, since it could apply to anything from a handful of siblings to, perhaps, the entire set of terrestrial organisms. Thus something more needs to be said about what makes a genealogically coherent set of organisms correspond to the rank of species.

A number of different answers have been given to this question. It is useful to consider the available options by comparison of the taxonomic hierarchy with a genealogical tree. The latter represents each divergence in the evolution of life with a branch. One, perhaps extreme, position, generally referred to as cladistic taxonomy or cladistics, proposes that taxonomy and genealogy should aim for complete convergence.[11] All and only distinct branches of the genealogical tree should be given distinct names. Proper taxa must be monophyletic; that is, they must include only the descendants of some ancestral branch of the genealogical tree. Strict cladistics requires also that taxa include *all* the descendants of that branch. Species are simply the smallest twigs of the tree. On a strict cladist view higher taxa are no less objectively real than species, since all are equally required to be monophyletic. To identify every branch in the tree, strict cladistics will require some very elaborate nomenclature. E. O. Wiley (1979) offers a suggestion for making things a bit more manageable.[12]

Probably a majority of those sympathetic to phylogenetic taxonomy are in fact committed only to the much weaker demand that classification not be inconsistent with the genealogical tree. Typically, this will involve the requirement that the members of a taxon be monophyletic, but not that the taxon necessarily includes all organisms descended from the relevant ancestral population. To take a standard example, the belief that birds are descended from (primitive) reptiles should, on strict cladistic grounds, commit us to classifying birds as a subgroup of the class Reptilia. The more tolerant and traditional classification of Reptilia and Aves as separate classes remains generally favored. In addition, the possibility of anagenetic speciation, of speciation within a continuing undivided lineage, need not be excluded; sufficient change within such a lineage may, to say the least, make the insistence on the same name for all its temporal parts extremely confusing. The general motivation for such divergence from strict cladistics is the thought that judgments of similarity and difference should have some relevance to taxonomy independent of the desirability of recording phylogeny. Such positions will thus require some appeal to criteria of speciation distinct from phylogenetic separation. Joel Cracraft (1983, 1987), for example, proposes the conjunction of one or more heritable diagnostic characters with the reproductive cohesiveness of the group; Leigh Van Valen (1976) defends an ecological criterion; and Gareth Nelson and Norman Platnick (1981) require a "minimal evolutionary novelty" (this position is generally referred to as "pattern cladism").

PLURALISTIC

The preceding observations lead naturally to the proposal of a moderate brand of pluralism. Subject to the constraint that species be coherent genealogical units, it has been suggested that we might adopt various different criteria for deciding to call such units species (Mishler and Donoghue, 1982; Donoghue, 1985). The crucial distinction at work is that between a criterion of grouping and a criterion of ranking (see Mishler and Brandon, 1987). A grouping criterion determines which organisms are even candidates for conspecificity, whereas a ranking criterion determines the actual extent of the species. This moderate pluralism insists on a monistic grouping criterion—monophyly—while allowing pluralism of ranking criteria: the latter may be selected according to what kinds of causal processes are of greatest relevance for maintaining the cohesion of a lineage in a particular case.

This move to pluralism reflects a fundamental difficulty with the phylogenetic approach to classification. The genealogy of life is enormously complex. It is likely, for example, that divisions can be found well below what is generally taken to be the species level, so that for a strict cladist the truly basal taxonomic units will often be quite small populations. (Indeed, it is not entirely clear how we should motivate any stopping point in constructing genealogies until we reach the individual organism.) So long as we ignore epistemological obstacles to the full disclosure of genealogy, this approach does provide the possibility of maximally unambiguous assignments of organisms to their evolutionary places; but it offers very little useful advice about *classification,* that is, about the breadth of categories that will be useful for various biological purposes. A natural response to this difficulty is to argue that questions about the breadth of classification should be addressed locally, in terms of either the particular features of the organisms in question or the particular problems that the classifier has in view. Thus within the constraint of consistency with genealogy, we may draw taxonomic distinctions on a variety of grounds, including methods of reproductive isolation, occupation of an ecological niche, or even a range of other morphological criteria selected in the light of a particular investigative goal.

The problem with this proposal is just that it is very difficult to see why criteria other than monophyly should provide classifications that fit onto the genealogical tree at all. Consider a very simple cladogram (or idealized bit of evolutionary history):

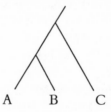

$$A \qquad B \qquad C$$

I can see no reason whatever why the group consisting of A and C, but excluding B, should not turn out to share the same ecological niche, or even share reproductive links. (These could, presumably, have been severed at the point where A and C divide, and reestablished at some time after the separation of A from B.) Thus, I cannot see why anyone who takes seriously the idea that different taxonomies may be needed for different theoretical purposes should nevertheless insist on constraining all such taxonomies to consistency with genealogy.

Given the insistence on this monistic grouping criterion, it is appropriate that B. D. Mishler and M. J. Donoghue refer to their view of species as (a variant of) the phylogenetic species concept. By contrast Kitcher (1984a) has advocated a radically pluralistic conception of species. Kitcher argues that both historical (evolutionary) and structural (or functional) inquiries should be accorded equal weight in biology, and that they may require different classificatory schemes, the latter in some cases demanding a morphological classification. This is the view that I wish to endorse. The schematic example in the last paragraph is the simplest possible case in which historical and functional classifications can diverge. In practice, there are likely to be much more complex cases. Nothing in evolutionary theory guarantees that genealogy will always provide us with the distinctions we need in order to understand the current *products* of evolution as opposed to the process by which they came to be. This suggestion is reinforced by the proposal that historical and functional classifications generate taxa with fundamentally distinct ontological statuses.

A different kind of motivation for this radical pluralism is provided by a possibility discussed above under the heading of the reproductive species concept: there is no metaphysical guarantee that whatever accounts for the coherence of species and their distinction (such as it is) from other species will be the same for all living things. The biological species concept already commits its devotees to denying that asexual organisms come within the scope of its general account of diversity. It can be no huge step to admit that among sexual organisms there may be a variety of mechanisms that explain and maintain biological diversity, and thus there may be organisms that form species, but not in the sense of the biological species concept.[13] Thus both ontological and pragmatic grounds support a pluralistic approach to the species question.

Somewhat more speculatively, I am inclined to suspect that the persistence and intractability of the species problem has much to do with a tension between the assumption that science is concerned with discovering the real and unique structure of nature (in the standard figure from carnivory, "carving nature at the joints") and the only slowly dawning realization that Darwin has bequeathed us a nature with no such unique structure. This insight should not, finally, be taken as necessitating complete tolerance of every taxonomic scheme that anyone should happen to think up. The point, rather, is to make it clearer what should be the grounds for accepting a taxonomic scheme: not that it is the right one, since there is none such; but that it serves some significant purpose better than the available alternatives.

One obvious and often-heard objection to this kind of pluralism is that it embraces conceptual nihilism and invites us into a taxonomic Tower of Babel. While it would certainly be nice if we could find product identification numbers stamped by God on the genes of every creature, the absence of such a wholly objective taxonomy is not nearly as threatening as proponents of this objection suggest. First, large-scale taxonomies are not the works of a single hand with one particular and perhaps esoteric taxonomic philosophy, but the compilation over time of the labors of many. Thus it is not so much a multiplicity of taxonomies reflecting different views of the species question that we should expect from pluralism, as taxonomies each of which may have many different sources, perhaps reflecting at times such ideologically divergent bases. It is not clear why this diversity should produce confusion, although no doubt the resource that such an evolving taxonomy provides will not meet the ideals of a taxonomic essentialist. Second, detailed local taxonomies may indeed be the work of a particular taxonomist with a particular theoretical perspective. But this is just where pluralism has the clearest benefits. The taxonomy may be constructed with particular goals in mind, goals that make a particular approach appropriate. Or it may be applied to a group of organisms for which one approach is particularly desirable or practically inevitable (for example, a morphological taxonomy of an asexual group). Since it is likely that those using such a taxonomy will be well aware what is going on, and even why, confusion is unlikely. And the benefits of allowing the people most familiar with a particular group of organism, and most aware of the central current problems concerning the organisms in that group, to determine the appropriate approach to classification seem almost certain to outweigh any residual danger of confusion.

A rather more sophisticated methodological objection to pluralism has recently been articulated by David Hull (1989). In defense of what he takes to be the currently best-supported, genealogical conception of species he admits the possibility that eventually "monism will have proved to have been only temporary" (pp. 120–121), but he insists that "the only way to find out how adequate a particular conception happens to be is to give it a run for its money. Remaining content with a variety of slightly or radically different species concepts might be admirably open minded and liberal, but it would be destructive of science" (p. 121). This, it seems to me, is plausible only if one is already committed to the view that science requires, in the end, a unified biology with a wholly univocal concept of the species. But the preceding arguments have claimed that far from being, or being likely to become, a unified project, biological science encompasses projects with various goals and addresses very disparate phenomena. Thus my motivation for pluralism is not methodological ("let a thousand flowers bloom, even the weediest ones"), but ontological. It is that the complexity and variety of the biological world is such that only a pluralistic approach is likely to prove adequate for its investigation. Certainly any successful scientific program should be given a run for its money. But I cannot see why science would be destroyed if no program is capable of finishing the course without assistance.

Species and Essentialism

It is sometimes argued that the reason that species cannot be treated as natural kinds is that they lack essences. A natural kind, on this conception, consists of all and only those things that share some essential property. It is very widely agreed among biologists that no such property can be found to demarcate species, so that if an essential property is necessary for a natural kind, species are not natural kinds. In the next chapter I shall develop a weaker account of natural kinds, which does not require that such kinds have essential properties. Here I shall briefly review the reasons for rejecting essentialism in biological classification. Possible candidates for essential properties will, of course, depend on one's conception of species. However, essentialism can be seen to fail regardless of whether one espouses the morphological, the biological, or the phylogenetic species concept. A pluralist, by denying that there is *any* uniquely correct scheme of classification, is clearly committed to the denial of essentialism.

An assumption that dates back at least to Aristotle is that organisms can be unambiguously sorted into discrete, nonoverlapping kinds on the basis of gross morphological properties. Since the theory of evolution undermined the belief in the fixity of species this position has become increasingly difficult to defend, and has indeed been almost universally rejected. Roughly, this is because almost all such properties show significant intraspecific variation, and the range of variation within a species will often overlap the range of variation of the same property within other species.[14] Moreover, even if this is not empirically the case, it is impossible to see how one could sustain the counterfactual claim necessary for taking such a property as an essence, that if such extreme variation were to occur, the organism exhibiting it could not be counted as a member of the species.

Nevertheless it is still sometimes thought that a more covert, probably microstructural, property could be discovered that would be adequate for the unambiguous assignment of individuals to species.[15] More specifically, it may be supposed that some description of the genetic material could provide a genuinely essential property.[16] Thus intraspecific differences might be taken as a consequence of varying interactions with the environment, allowing the possibility that all the members of the species share the same genetic blueprint, or one with certain common essential features. But it is equally possible that intraspecific genetic variability might overlap interspecific variation as much as, or more than, morphological variability does. In fact there are good reasons for supposing this to be the case.[17]

A number of reasons have been suggested why natural selection might tend to generate species with a high degree of genetic variation. First, a reserve of genetic variation may enable the species to survive changing environmental conditions, by producing individuals adapted to a wide variety of circumstances. Second, it has been argued that heterozygous individuals (that is, individuals with pairs of different genes at various loci) may be better adapted than homozygous ones. The classic exemplar of this possibility is human sickle-cell anaemia. Generally, it is plausible that heterozygosity may usefully expand the biochemical resources of an individual. Third, as was discussed in the previous section, it is believed that there are homeostatic developmental mechanisms whereby differing gene combinations can approximate the production of the same phenotype.[18] This last point provides a very concrete illustration of the possibility that genetic variation might exceed morphological, and emphasizes the baselessness of the expectation that genetic properties should provide any exception to the general post-Darwinian emphasis on variability.

It might reasonably be asked here whether these epigenetic mechanisms might not themselves serve as essential properties. And I think that if, as I speculated earlier, there are species for which these provide the best account of species coherence, we would have here perhaps the best candidates in biology for real essences. This would, however, constitute a very modest concession to essentialism. First, I have not suggested that this criterion of species membership would be adequate to most or even many biological projects. And second, it is an empirical question what proportion of kinds of organisms have extremely invariant such mechanisms, or whether many might have epigenetic mechanisms allowing variation comparable to any other aspect of biological morphology. (This, of course, might be the case for species whose coherence was maintained by intraspecific gene flow.) Thus while epigenetic mechanisms might serve most or all of the functions of a real essence for some kinds of organisms, they are unlikely to be at all generally available. And, contradicting one of the main points of essentialism, there is no reason in principle why the species should not survive the demise of these mechanisms (some other cause of species coherence gradually taking over). At any rate, a truly tolerant pluralism should surely allow the occasional appearance even of a plausible candidate for a real essence.

The evolutionarily based accounts of the species concept are on the whole even less encouraging for essentialism. Presumably on the biological species concept, the only candidate for an essential property would have to be grounded in reproductive isolation: the capacity to form reproductive links with all but only members of the same species, or perhaps the physical basis of that capacity. One obvious limitation parallels a limitation to the biological species concept: it has no possible application to asexual species. A more fundamental problem is that reproductive isolation is by no means always absolute. Hybridization, in fact, occurs throughout the biological world, although it is especially prevalent among plants, fishes, and amphibians. This is not necessarily an insuperable problem for the biological species concept, as there are many ways in which introgression of alien genes can be sufficiently limited to maintain the integrity of a species without requiring absolute barriers to interspecific reproduction. Hybrids may be sufficiently competitively inferior to nonhybrids that natural selection can maintain this integrity, for example. But hybridization does provide an insuperable obstacle to seeing reproductive capacity as grounding an essential property of a species. First, there will be individuals that will not be assigned to any species on this criterion; and second, there will be reproductive links between individuals that certainly belong to different species. The latter point is reinforced by the observation that the ability to produce viable offspring is not transitive. There exist chains of species, so-called ring species, any two adjacent members of which can produce viable offspring but the terminal members of which are unable to interbreed. Whereas these phenomena provide significant but not necessarily insurmountable difficulties for the biological species concept, they are devastating objections to any attempt to ally this species concept with essentialism.

Last, and least promising for the essentialist, is the phylogenetic species concept. To begin with, the only possible essential property derivable from this conception would be purely relational—the property of being descended from a certain group of ancestors—and it is not clear whether this is even a candidate for an essential property.[19] I am inclined to deny that such a property could be essential, by appeal to a simple thought experiment. If, say, a chicken were to lay perfectly ordinary walnuts, which were then planted and grew into walnut trees, I would be disinclined to refer to this as the production of a grove of chickens. If this intuition is accepted, it shows that the right ancestry is not a sufficient condition for species membership. My intuition, moreover, is that the trees in question might prove to be genuine walnut trees, which is to deny that ancestry is even a necessary condition. However, I do not want to put much weight on this kind of argument, and I think there is a more substantial objection to the proposal under view. Any sorting principle based on ancestry supposes that it is possible in some way to pick out the appropriate set of ancestors. In other words, being descended from one of the members of a particular set is no criterion at all unless there is some way of picking out the members of that set. Of course one could push the question back to some other set, but unless we are prepared to reduce classification to the view that we are all members of one huge taxon, the descendants of the first organisms in the primeval slime, at some point we must appeal to more concrete features. Thus unless some set of features independent of ancestral relations were available to sort the putative ancestors unambiguously and exhaustively, the descendants could not hope to claim such a perfect classification. But I have already argued that this precondition cannot, in general, be met. Phylogeny, in short, cannot possibly create essences *ex nihilo*.

This exposition of some of the reasons why essentialism has been so widely rejected in application to biological classification extends the thesis of the previous chapter into the scientific domain. There is no God-given, unique way to classify the innumerable and diverse products of the evolutionary process. There are many plausible and defensible ways of doing so, and the best way of doing so will depend on both the purposes of the classification and the peculiarities of the organisms in question, whether those purposes belong to what is traditionally considered part of science or part of ordinary life.

One final question must be addressed: Is the kind of pluralism I have been advocating consistent with a realistic attitude to the various kinds, and even individuals, that I have discussed? There are a number of pluralistic possibilities that I have defended, but none, as far as I can see, forces one to abandon realism. Least problematic is the idea that different parts of the taxonomic tree might be segmented according to different criteria of division. Realism about biological kinds has nothing to do with insisting that there should be some unitary cause of biological distinctions. Provided realism is separated from certain essentialist theses, I see little more reason why the possibility of distinct and perhaps overlapping kinds should threaten the reality of those kinds. Just as a particular tree might be an instance of a certain genus (say

Thuja) and also a kind of timber (cedar) despite the fact that these kinds are only partially overlapping, so an organism might belong to both one kind defined by a genealogical taxonomy and another defined by an ecologically driven taxonomy. The only obvious precondition of this possibility is that kinds not be so homogeneous that membership of a particular taxon could be sufficient to determine everything significant about an organism. And this kind of heterogeneity is universally conceded. I do not take these remarks to bear on the more abstract debate between realism and nominalism. But if one concedes the possibility of being a realist at all, I cannot see why one should balk at promiscuous realism.

It is perhaps less clear how a species could be both a (real) individual and a kind. To some degree the plausibility of this idea depends on that of a pluralism of kinds of kinds. For clearly those kinds that are defined in terms of intrinsic rather than relational properties (that is, genealogy or reproductive links) are not even candidates for comprising individuals. Might a kind defined in terms of, say, reproductive isolation be both an individual and a kind? Strictly, the answer must be no, since kinds and individuals are objects of wholly different ontological categories, and certainly nothing could belong to both these categories. But this only forces us to state the possibility in question with greater precision, and this, I believe, quickly disposes of the difficulty. The real question is whether the same set of individuals can provide both the extension of a kind and the constituent parts of a larger individual. And the answer to this is clearly yes, as illustrated by the example of myself, and the set of cells with my particular genotype. There are difficult philosophical questions about the nature of individuals that may cause problems for the claim that species are individuals, and that are beyond the scope of the present discussion. But I do not think that the fact that many species are also kinds presents such a problem.

This development of the idea of pluralistic, or promiscuous, realism has shown that what was found true in the preceding chapter of the categories of everyday life is as true of scientific attempts to classify the same range of phenomena. It has also introduced the issue of essentialism in science. In the next chapter I shall present a more general account of antiessentialist metaphysics and suggest that there is some use to the concept of a natural kind, provided it is shorn of its essentialist trappings. In addition, I shall take the antiessentialist argument one stage further by considering a far more general and abstract category than any particular taxonomic kind of organism, sex.

3. Essences

It is still widely believed that science is, among other things, the search for fundamental kinds defined by real essences. In the preceding chapters, on the other hand, I have presented a picture of biological phenomena that highlights the great variety of ways in which the biological domain can be divided up, and have tried to show that there is nothing amiss with such a pluralism of classifications. This picture is directly at odds with a belief in Lockean real essences; for the existence of such real essences would imply that there is some unique, privileged scheme of classification, which assigns everything to a class defined by common possession of the appropriate essence. While the existence of such a privileged scheme might be compatible with the existence of disparate categories for the rough-and-ready purposes of everyday life, it surely does entail that only the one privileged scheme is adequate for the purposes of science. In the last chapter I argued that there is no place for such an essentialist taxonomy even in the scientific classification of biological organisms. A possible response to this begins with the concession that groups of similar or related organisms do not form any fundamental kinds, or even, according to the Ghiselin/Hull thesis, kinds at all. The search for true natural kinds with real essences, however, could be carried on elsewhere, presumably at some higher level of abstraction. In this chapter I shall present some more general arguments against the idea that science should be committed to the search for real essences. I shall illustrate the argument by appeal to a promising candidate for such a higher-level natural kind, sex. By arguing that sex is indeed an indispensable concept for biology, but also that it admits no possibility of an essentialist interpretation, I hope to support the idea that natural kinds themselves should be reconceived in a nonessentialist way.

The main thrust of my argument against essentialism might be described as a plea for categorial empiricism. Unless one supposes that the delineation or discovery of a kind implies the discovery of an essence, there is no more to the discovery of a kind than the discovery of the correlations of properties characteristic of the members of a kind. If one did suppose that one had at the same time as discovering the kind also discovered the essence of the kind, one would thereby have additionally determined the extension of the kind. The extension, that is to say, would simply be the set of things that shared the essence. By "categorial empiricism" I mean the view that discovering the extension of a kind, and thus the range of application of an empirical generalization, is not that easy. This point is forcefully illustrated by the concept of sex. For although this is undoubtedly a concept that has major significance for biology, and although it is also a concept that divides the natural world into well-defined classes, the scope of the generalizations to which it gives rise is at every stage an empirical matter, and varies from case to case. This point will be developed further with application to the concept of gender, which encompasses the ramifications of sexual dimorphism in the context of human society. In this case a great deal of detailed scholarship has documented the dangers inherent in simplistic assumptions about the range of generalizations. Perhaps the most general conclusion from this discussion is that the search for kinds, as opposed to the exploration of properties and their correlations, has been overestimated as a component of scientific inquiry.

In the following pages I shall first discuss my understanding of essentialism in more detail, and begin to amplify the antiessentialist argument I have just sketched. I shall then discuss in some detail the concepts first of sex, and then of gender, and return at the end of the chapter to a final elaboration of categorial empiricism.

Grades of Essentialism

In thinking about essences, it is first necessary to distinguish two very different functions that they may be supposed to serve. First, essences are often conceived as properties that determine the answer to the question to what kind the objects that instantiate them belong. If E is the essential property of kind K, and an object O has E, then K is *the* answer to the question, What kind of thing is O? But, second, essences are also thought of as determining the properties and behavior of objects belonging to the kinds of which they are the essences.

Considering the first kind of function, it is useful to recall the Lockean distinction between real and nominal essences. The view that nominal essences determine the kinds to which objects belong amounts to little more than the introduction of a bit of technical terminology. A nominal essence is connected to a kind by some sort of linguistic convention. Since it is obvious that we could not refer generically to members of any kind without the existence of some linguistic convention determining the (at least approximate) intended extension of the kind, the existence of nominal essences as just characterized is not controversial. More detailed descriptions of nominal essences may certainly commit one to more or less powerful semantic theses—for example, to the view that there must be necessary and sufficient conditions for the application of every classificatory term—but will carry no metaphysical commitments.[1] My concern here will again be solely with real essences.

To assert that there are real essences is, in part, to claim that there are fundamental properties that determine the existence and extensions of kinds that instantiate them. The existence of such properties would have profound metaphysical consequences. In particular it would imply that the existence of kinds of things was as much a matter of fact about the world as was the existence of particular things. Such kinds would be quite independent of our attempts to distinguish them, and their discovery would be an integral part of the agenda of science. The majority of contemporary usage takes such independent existence of kinds defined by real essences and awaiting scientific discovery as constituting the necessary and sufficient condition for the existence of a *natural* kind. Certainly, the existence of a natural kind (ignoring possible problems about noninstantiation) will follow from the existence of a real essence. I do, however, wish to dissent from the thesis that a real essence is *necessary* for a natural kind. I believe that it is quite appropriate to describe many of the ordinary language kinds discussed in Chapter 1 as *natural* kinds. If cedars, for example, do all serve a common function for the carpenter, then relative to that human practice I see no reason to deny the naturalness of the kind they form: it is a natural fact, after all, that there exists a class of things, albeit botanically diverse, suited to this role. I see no reason to deny that this kind might exist independently of our recognition of it, and might have been literally discovered by early carpenters. Evidently, on this conception the naturalness of kinds will turn out to be a matter of degree: some kinds will be a good deal more natural than others. I have no serious quarrel if someone wishes to reserve the term *natural kind* for a much narrower and more precisely delimited class of kinds that do share an essence. The motivation behind this insistence, however, will depend on how prevalent such natural kinds (with essences) turn out to be. Believing them to be few and far between, I prefer the more tolerant usage. In view of this preference, I shall use the term *strong natural kinds* to refer to natural kinds assumed to share a common real essence. When I refer simply to *natural kinds,* I shall not intend an essentialist interpretation.

Crucial for my argument against essentialism is the observation that even if a kind *is* determined by a real essence, the *discovery* of such an essence presupposes the discovery of the kind. Only the most extreme reductionist could suppose that examining a particular individual would allow one to determine to what kind it belonged apart from the prior recognition and at least partial characterization of that kind. This simple observation should be sufficient to raise serious doubts about the empirical credentials of real essences, especially to the extent that substantial consequences are alleged to follow from the presence of an essence. As I shall argue, if the existence of a real essence amounts to anything, it does, indeed, have practical consequences—specifically, in entitling us to anticipate the existence of laws governing the behavior of objects that partake of it. The subsequent discussions of sex and gender will show how misleading any such anticipations can be, even in cases in which there is no question that we are considering a significant natural category. (In my tolerant sense sex, surely, defines legitimate natural kinds.) The conclusion is that discovering kinds does not involve discovering essences; and so, given that there is no other way of discovering them, nothing does.

It does seem, nevertheless, that, barring the most radical and implausible nominalism, there should be something to the doctrine of real essences as so far described. Some kinds are more natural than others. The class of creatures with wings and feathers, for example, is more natural than that of creatures that are gray and over one foot long. This is so because when we know that a creature belongs to the first class, we can make numerous further predictions about it: that it or its female counterpart lays eggs, is warm-blooded, relatively large-brained, and so on. Membership in the second class carries no such benefits. Depending how deeply we can explain such clustering of features, we can adduce more or less powerful characterizations of essences.[2] If, to take one extreme case, God had simply chosen to assemble creatures in the light of some preconceived ideas about which features went well together, these essences might amount to no more than conjunctive, or perhaps partially disjunctive, descriptions of God's aesthetic preferences. Since these descriptions would still reflect genuine clusterings of properties, they would be natural kinds and might even be said to possess a kind of real essence. However, somewhat deeper explanations are reasonably expected for such discontinuities in nature. And this leads us to the second, and more problematic, function of real essences, that of explaining the nature and properties of the members of the kinds that such essences are supposed to determine.

The strongest possible notion of a real essence would be that of a property, or group of properties, that determined—and hence, in principle could be used to explain—all the other properties and behaviors of the objects possessing them. Although such a notion *might* be defensible for individual essences (Locke seems sometimes to have envisaged a microstructural description of an object potentially playing this role), it cannot work for the type-essences that are my present concern, for the simple reason that there are no kinds (with the possible exception of the ultimate microphysical kinds) the members of which are identical with respect to *all* their properties, even their intrinsic ones. I emphasize *intrinsic* properties because it is obvious that the *behavior* of an object will typically depend on both its intrinsic properties and its external environment. Clearly, the strong view I am considering should claim only that essence determines behavior as a function of some set of external variables or, in other words, determines precisely specifiable dispositions to behavior. However, variation in intrinsic properties and dispositions requires a more fundamental retreat from the strong position. Specifically, some distinction between essential and accidental properties, that is, between properties that can and those that cannot vary between members of a kind, is unavoidable. ("Essential" here includes both properties constitutive of the essence and properties fully determined by the essence.)

A promising and natural modification of the strongest conception of real essence, which provides a way of drawing just the distinction mentioned above, is the following: the essence of a kind determines just those properties and dispositions of its instances for which it is a matter of natural law that members of the kind will exhibit those properties or dispositions. Thus the real essence of a kind will nomologically determine some range of further properties of its members; these properties are (derivatively) essential. Those properties of an individual member of the kind for which this is not the case are merely accidental. For example, it is clearly no law of nature that squirrels are gray, since many are black. On the other hand, perhaps it is a law that squirrels have tails and, hence, tailedness is an essential property of squirrels. An essentialist holding the position I am now considering would explain this by saying that the essence of squirreldom, perhaps a particular genetic structure, determined the growth of tails but not a particular color of coat. The suggestion that the genetic structure might be the essence illustrates a further important aspect of such a position, the way in which the essence itself is to be distinguished from other essential properties. Presumably, the genetic structure causally determines the growth of a tail, and not vice versa. Thus, the essence itself is that property or set of properties that is explanatorily primary among the set of essential properties. Perhaps more precisely, one might suggest that the essence consisted of the smallest set of properties from which the entire set of essential properties could be inferred.

Making such an essentialism applicable to any part of biology—and, probably, to most other parts of science—requires some significant qualifications. To begin with, it is very difficult to find really sharp distinctions anywhere in biology; generally, there is a range of intermediate cases. Certainly, as far as taxonomic distinctions are concerned, sharp boundaries are the exception rather than the rule. Insofar as candidates for essential properties fail to distinguish sharply between the haves and the have-nots, a theory of essences would have to be considered as applying to typical members of kinds rather than to all members (unless some individuals are excluded from membership of *any* kind). Assuming that it does remain desirable to attribute individuals to kinds despite their abnormalities, the laws applying to such kinds could be only probabilistic. The probability that something has a tail, given that it is a squirrel, would then reflect the frequency of the abnormality of tailedness. The modified essentialist position could be maintained by insisting that there is, nevertheless, some standard genetic structure that constitutes the essence of squirreldom, and that anything that realizes this structure will, as a matter of nomic necessity (and barring extreme abuse), have a tail. Less-than-ideal squirrels would then be judged to be squirrels, or not, on the degree of similarity of their genetic structure to this standard form. It will be apparent, however, that this unavoidable modification leaves the essential/accidental distinction disturbingly arbitrary. It is, for example, unclear whence a fundamental distinction between blackness and taillessness of squirrels could derive—apart from a patently question-begging appeal to the essential nature. If it comes to no more than a quantitative difference in frequency, then an arbitrary decision is required to include one (or its genetic basis) but not the other.

Many philosophers of biology will be inclined to object at this point that the examples I have been considering are quite inappropriate since species are either not kinds at all (as discussed in the

previous chapter), or if they are, they are very far from fundamental kinds. Although I believe that species are, at least in some
contexts, kinds, I do not think the issue turns in any way on this
question. The example can be treated as merely illustrative. It is
true that fundamental biological laws are generally hypothesized
with much more abstract and general kinds. But species, populations, interactors, avatars, predators, or whatever are surely far
less homogeneous than squirrels, and at least as far from exhibiting
absolutely determinate extensions. The example of sex, to be taken
up momentarily, illustrates the difficulties that arise in attempting
an essentialist understanding of a more theoretically interesting
concept. Surely male and female constitute a genuine and important distinction in biology. Yet in sexually dimorphic species there
are typically variations with respect to sexually dimorphic characteristics, and even genuinely intermediate individuals. Moreover, the elaboration of sexual dimorphism varies greatly from
one species to another. Since the distinction between fairly sharp
boundaries and absolutely sharp ones is itself an absolutely sharp
one, the advocate of biological kinds completely immune to the
present problem will have a difficult task.

This difficulty highlights the detachability of a belief in natural
kinds from a belief in real essences. The belief that there are
discontinuities in nature to be discovered rather than invented is
quite independent of the question whether these are sharp or
gradual. Moreover, the importance of natural kinds for explanation does not depend on a doctrine of essences. One might imagine, for example, that there was some optimal set of laws (perhaps
maximally deterministic and/or explanatory) describing a particular domain, and that the classes of entities referred to by those
laws should be considered as natural kinds. Such a view requires
no fundamental distinction between the essential and accidental
properties of members of such kinds. Since it may well be the
case that the frequencies of properties in a kind will vary continuously from near universality to nearly total absence, such a distinction can seem profoundly arbitrary. But, as I have tried to
show, without such a distinction the point of essentialism becomes
obscure.

Sex

As plausible an essential difference as one is likely to find in biology is that between male and female. The distinction can be drawn for a very large number of kinds of organisms, and although, as with almost all biological distinctions, there are borderline cases, the vast majority of organisms of types to which the distinction applies can be assigned unambiguously to one category or the other.

Another relevant feature of the distinction, though now only for as long as we look at a particular type of organism, is that there are systematic differences between males and females at various levels of structural organization, and that these are causally and explanatorily related. More specifically, in most species males and females differ genetically, physiologically, and behaviorally; and it is believed that the genetic differences causally influence the physiological ones, and that the physiological differences influence the behavioral ones.

However, further consideration shows that this situation diverges greatly from the essentialist scenario sketched in the previous section. The properties that are causally fundamental in explaining sexual dimorphism between the members of a species are unquestionably not those that constitute the real essence (if any) of maleness and femaleness. A microstructurally oriented essentialist might be inclined rashly to assume that the essence of maleness and femaleness for humans was the possession, respectively, of an XX or XY chromosome. But many animals that can be divided into males and females as clearly as humans can have no XX or XY chromosomes. This view would imply that to say that there are both human females and female geese would be an equivocation on the word *female,* since in each case the word refers to a quite different microstructural property. And this would be patently absurd.

Surely the correct way of describing the situation is to say that *for humans,* having XY or XX chromosomes *causes* individuals to be male or female. What it is to be male or female, on the other hand, is a property at a much higher level of organization, that of producing relatively large or small gametes. It is *this* distinction, based on the fact that most types of organisms have individuals of two kinds distinguishable by a major dimorphism in the size of gametes they produce,[3] that is referred to by the general categories of male and female, and that in particular species is caused by a particular genetic dimorphism. This leads to the surprising conclusion that physiological differences between the sexes, and whatever genetically determined behavioral differences there may be, are not, in fact, caused by the sex of the organism; rather, both the differences and the sex are joint effects of a common cause, the sex-determining genetic structure.

In this light, it is less surprising that the sexual categories prove to have rather little explanatory power. It is very doubtful whether there are any very significant laws relating to males and females in general. It is possible and even likely that every generalization about a sexually specific characteristic is limited to some range narrower than that of all sexually dimorphic species. In some cases there is an identifiable taxon over which the generalization applies, either because the character is an evolutionary novelty in a phylogenetically distinct taxon, or because that very characteristic is used to define a higher grouping, as with mammals or placental mammals (whether these are distinct possibilities will depend on one's general theory of taxonomy).

Although the possibility cannot be ruled out a priori that there might be some properties universally, or almost universally, correlated with large or small gamete production, there seems little reason to expect that this will be the case. This observation may invite reconsideration of my claim that sexual categories are outstanding candidates for biological natural kinds. The intuitive basis for that claim certainly has nothing to do with a knowledge of laws pertaining to males and females in general. It is based, rather, on two kinds of observation. First, for enormous numbers of species, a generally sharp distinction can be found between males and females. And second, within any sexual species, and often within much larger taxa, there are pervasive sex-specific generalizations to be made. Men grow (or shave) beards, and women grow breasts; males and females of many large groups of species have relatively similar genitalia; and so on. However, what these observations properly suggest is that sex is a very significant property that may be appealed to in the analysis of numerous different taxonomic groupings but that, nevertheless, it is not a property that is sufficient to define any significant kind. Alternatively, if one wishes to insist that males and females *do* form kinds, then there are natural kinds with little explanatory power.

The fact that nature can be "carved at the joints" without yielding explanatorily significant categories is of great significance. The explanation in this case is not hard to find, deriving from a pervasive fact about biology: biological kinds reflect historical similarities as much as they indicate similarities of causal power.[4] The divide between males and females, as general categories, derives not from characteristic properties or dispositions of the two classes but, presumably, from a powerful evolutionary tendency toward sexual dimorphism.[5] But it is quite possible, even likely, that this common evolutionary tendency may do no more than favor a simple dimorphism of gamete size, and that subsequent elaborations of the dimorphism may be quite specific to particular evolutionary lineages, and thus not susceptible to large-scale generalizations.

Two responses to the preceding argument need to be considered. First, I have so far ignored a trend in contemporary biology that *does* want to maintain the general explanatory power of sexual categories. Here I refer to a large area of sociobiology. And, second, one may accept the general conclusion that I have defended above but explore the possibility of defining narrower, but still explanatorily powerful, sexually delimited kinds.

I cannot hope to give an adequate treatment here of the highly problematic and controversial topic of sociobiology.[6] However, a major sociobiological thesis does assert precisely what I have said there is no reason to believe: that the simple fact of gamete size dimorphism strongly disposes species to certain subsequent evolutionary developments, specifically, to quite predictable behavioral dimorphism. At its most general, the theory asserts that those organisms with smaller gametes (that is, males) will tend to develop behavioral strategies that maximize the dispersion of their gametes, while the females will tend to develop strategies that increase the chances of successful maturation of those offspring that they are able to assist. At this very general level, the theory is based simply on the idea that a large gamete is a more significant investment of resources than a small one, and such a prior commitment will give disproportionate encouragement to strategies that help preserve the investment. Sperms, on the other hand, are a dime a dozen, and it is therefore more profitable for the male to invest his energies in distributing as many as possible than to worry about individual projects. If there is any force to this argument, there is apparently a lot more to it when the reproductive physiology of the organism requires that very large investments of resources are demanded of the female to have any chance of successful reproduction, as is the case with viviparous animals or those that lay large eggs. Finally, in such cases, it is argued that when the offspring, or egg, is produced a substantial time after fertilization and requires further care to have any chance of survival, the female will find herself playing an evolutionary game with no cards. The male, it is argued, will by then have taken off on a quest to impregnate more females, confident in the (implicit, evolutionary) knowledge that the only way the female can expect to have any reproductive success at all will be to provide at least the essential minimum of parental care. (It has not escaped notice that this story recreates and naturalizes the traditional human stereotypes of the fickle promiscuous male and the reliable, nurturing female. This issue will be taken up in the next section.)

Despite the initial plausibility of this argument, it suffers from the serious defect that the predictions to which it gives rise do not turn out to be true. Many species, even of birds and mammals, are quite monogamous in both sexes, and there are many species in which the male provides as much parental care as the female, or even more. But I shall not attempt any evaluation of the general worth of the sociobiological argument, since the simple observation of variation in sexually specific behavior across species is sufficient to draw the conclusion relevant to my present purposes. This is simply that however significant a *force* in evolution these considerations show gamete size dimorphism to be, that is all they show. Clearly, if there is such a force, it is one capable of being overridden by other opposing forces; otherwise, there could not exist the many exceptions just referred to. (The same point can, and will, be made in connection with alleged systematic behavioral differences between men and women.) It might be thought that since I have allowed that dispositions common to members of a kind suffice to give that kind explanatory power, the above concession should be sufficient to constitute sexual categories as natural kinds. But this would be a confusion based on the failure to distinguish historical from causally explanatory categories. It may be that in every species there has been an evolutionary tendency for males to acquire dispositions to promiscuity and females to acquire dispositions to parental care. But in many cases those tendencies have not been actualized; hence, the present members of many species do not have those dispositions. A drake, perhaps, has no disposition to desert his mate, or a female stickleback to care for her young. And it would be absurd to say that the drake must have such a disposition merely on the grounds that his ancestors had some, in fact unrealized, tendency to evolve such dispositions. So, in short, whatever the value of these sociobiological arguments, and although they may help to explain the particular behavioral dimorphisms in particular species, they do nothing to make males or females generally into genuinely explanatory kinds.[7]

The second response to which I alluded above was to accept that sexual categories are not themselves explanatory kinds but to argue that more narrowly defined sexually specific categories might, nevertheless, be such. Thus, male and female mammal, goose and gander, or man and woman may constitute sex-specific natural kinds with explanatory potential regardless of whether male and female are themselves such kinds. Two general points should be made about this proposal. First, assuming—which is highly dubious for any taxonomic level above the species—that the taxon that is being sexually restricted is a genuine kind, this is not the case of the intersection of two kinds, but one of the subdivision of one kind.[8] This point follows simply from the main conclusion just reached about general sexual categories.

There is nothing obviously problematic about such a nesting of natural kinds. Metal and iron provide one plausible example. Perhaps human and woman provide another. There is nothing incoherent, though almost certainly something false, in conceiving biological taxonomy this way. One qualification must be registered. It would be absurd to suppose that man and woman, say, were "better" kinds than human. To admit that species are kinds is to admit that kinds may encompass very considerable variation, and hence may license only probabilistic nomological generalization. Moreover, it is to admit that kinds need not be defined by any kind of real essence. It would, again, be absurd to suppose that the essence of woman was any easier to find than the essence of human. Since finding the former presupposes finding the latter, the reverse must be the case.

My second point follows, once again, from the fact that male and female are not themselves explanatory kinds. The explanatory significance of sexually specific kinds must be wholly empirically determined. No systematic differences between the males and females of a particular species can be assumed beyond those that are used to distinguish the sexes. This does require one qualification. Sometimes one can appeal to higher-level generalization. If one discovers a new species of mammal, one will reasonably anticipate that dissection will reveal a familiar and sexually dimorphic reproductive physiology typical of mammals. Generalizations may be empirically discoverable at various levels, and physiological generalizations will often have quite broad scope. Variation in behavior, on the other hand, tends to be much more fine-grained. Though not invariably or necessarily illegitimate, inference from one species to a related species will be much less secure. At any rate, whatever more or less empirically warranted inferences there may be within relatively phylogenetically narrow groups, this is quite different from attempting to deduce behavior or even morphology from mere subsumption under the broadest sexual categories. Accepting, at least for the moment, the possibility, if highly qualified, of sexually specific groups of particular kinds of organisms, it is time to look in more detail at the human case, and turn to the topic of gender.

Gender

The term *gender* has been developed in its present sense out of the last twenty years of intensive feminist scholarship, and refers to the sexually specific roles occupied by men and women in various societies. The reason that the distinction between sex and gender has been considered so crucial[9] is that whereas whatever properties follow solely from the sex of their bearers, such as reproductive physiology and secondary sexual characteristics, should be equally prevalent in all societies,[10] it is quite clear that gender roles, on the contrary, are highly variable and culturally specific in many respects. So even if man and woman as biological categories are modestly explanatory natural kinds, it is clear that much of the behavior encompassed by gender roles lies outside their explanatory scope.

It is not altogether easy to assess the *extent* of variability in gender roles. One reason for this is that a great deal of apparently relevant research is particularly susceptible to the kinds of problems to which feminist critics of science have drawn attention; if there is any part of science for which the accusation of distortion seems particularly plausible, this is surely it.[11] But both anthropological and historical evidence leaves little doubt that such variability is significant and widespread.[12] Of particular interest is such research on areas of behavior that have been especially prominent in attempts to reduce gender-specific behavior to a causal consequence of physiologically dimorphic features (genetics, hormones, brain structures, and so on). Promiscuity and the extent to which it is a male prerogative provide one example. Also important is the variability in the extent to which the generally socially sanctioned gender-specific behavior is adhered to or insisted upon. The prevalence of, and attitudes toward, homosexuality and incest, both subjects that have received a good deal of attention from sociobiologists, appear also to be highly variable. The weight of evidence seems to favor the opinion that in most, if not all, of the aspects of behavior that suggest sexual dimorphism in the context of a particular culture, there is a great deal of cross-cultural variation.

A traditional essentialist view of this issue might be the following. Both humans and males constitute natural kinds with some essential property. To be a male human is to partake of both the relevant essential properties, and much of the behavior of human males (aggression, rape, promiscuity, and so on) can be explained by reference to one or both of these essential properties. Against this I have argued that the most that can be sustained is the claim that male humans form a subkind of humankind. If this kind has an essential property it is presumably a combination of the essential genetic structure of humans with the specifically sex-determining genetic features of male humans. The reason that the essentialist is forced into this specific and rather unpromising form of genetic determinism is just that maleness in general is not an explanatory category, and the only available candidate for an explanation-generating essence for the kind of human males is their distinctive genetic features. Unpromising or not, there are certainly those, certain sociobiologists providing their theoretical wing, who want to maintain a position of this kind and trace the behavioral differences between men and women to the genetic.

Of course, all that the general conclusions about broad sexual categories show is that there is no general argument grounded in the essential features of a putative strong natural kind, maleness, that legitimates this kind of explanatory strategy. And this does nothing to show that the strategy may not nevertheless succeed. Moreover, it is surely plausible enough that in many species it will be possible to find genetically grounded sexually dimorphic patterns of behavior, regardless of any general thesis about maleness and femaleness. If this is possible for scorpion flies and ducks,[13] why not for humans? The core of the answer has already been indicated, in the brief discussion above of human cultural variability. After all, a thesis about the genetic determination of some behavior in any species would be deeply threatened by the discovery that the behavior in question varied greatly among members of that species. Given that behavioral variability seems so much more characteristic of humans than of any other species, it is important to consider how deep this difference cuts. It has become very unfashionable to claim fundamental divides between human and other animal natures. Nevertheless, I want to suggest that human cultural variation does provide a major differentiating feature of the human species, at least with respect to the matter of present concern, the relation between classification and explanation.

It is sometimes remarked that culture is merely a biological feature like any other, something that evolved to an extreme degree in one particular lineage of primates, and thus nothing that could seriously warrant radically different treatment from other biological features. However, it has also been noted that culture provides the possibility for processes analogous to, but distinct from, biological evolution to occur within particular societies. Such processes have been termed "cultural evolution."[14] Like features of organisms in a species, behavior in a society is not entirely uniform. The processes by which the behavioral repertoire of a society is transmitted from one generation to the next may well include selective mechanisms that permit systematic changes over time. For example, patterns or styles of behavior that are perceived as being, in some sense, particularly effective may be transmitted more frequently. This feature of cultural evolution, incidentally, suggests an important difference between biological and cultural evolution, that the latter may be directed toward identifiable goals.

Recognition of the possibility of cultural evolution raises a further possibility especially germane to the present topic. Biological evolution, as the title of Darwin's most famous work reminds us, was originally intended as a theory of biological diversity. The parallel between biological evolution as explanation of the divergence of groups of organisms and cultural evolution as a general account of the human social divergence seems to me compelling. Societies do change over time, and to a significant degree developments in one culture can be isolated from other cultures. A range of possible isolating mechanisms might include preferential transmission from parents or particular teachers; deliberate attempts to preserve cultural identity; prejudice; and, in some remaining cases, geographic separation. As in the biological case, isolation does not need to be total. The upshot of this is that for certain purposes it may be useful to think of cultural evolution as generating a variety of cultural species analogous to the biological species produced by Darwinian evolution.[15] This suggestion, finally, provides a somewhat nonstandard rationale for a claim that from certain more humanistic perspectives may seem banal: for understanding and explaining behavior, the human species is much too broad a category to which to appeal. Regularities in behavior often apply within much smaller categories, such as cultures, societies, or what I have called cultural species. The point is a further illustration of an idea to which I have constantly returned: the appropriate ranges of generalizations are enormously variable from case to case, and must always be seen as a purely empirical matter.

The implications of the preceding discussion for the case of gender-specific behavior should be clear. The feminist research on the historical and anthropological variability of such behavior illustrates that it is indeed an area of human behavior in which the scope of generalizations is typically limited, and less than the entire human species. Although this variation does not show that genetics plays no role in such behavior—indeed, it is unclear what it would mean to say that genetics played *no* role in *any* behavior— it does show that unifactorial genetic explanations are generally if not always inadequate.

Feminist theorists have also proposed alternative schemes of explanation for gender-specific behavior, most notably as largely epiphenomenal to the general political and economic subordination of women. If there does appear to be a clear cross-cultural universal in this area, it is that men seem almost invariably to have achieved a position of domination over women. But although this may provide an explanation of much gender-differentiated behavior, it is itself in serious need of explanation. Providing this explanation is one of the central theoretical goals of feminism. Awareness of this problem is also a large part of the reason why a significant component of feminist scholarship does not fit neatly into the general project I have so far indicated. A few feminists pretty much accept the essentialist structure I have been describing, while objecting only that the details have generally been filled out in a way revealing deep male bias.[16] Perhaps more significantly, a markedly essentialist flavor has been detected in a good deal of much more mainstream feminist thought.[17] Indeed, it may even seem that the very intelligibility of feminism depends on construing women as a strong natural kind and, hence, on accepting essentialism.[18]

An observation that does much to put this situation in the right perspective is that feminism is not merely a theoretical movement but, perhaps even more centrally, a political movement. If male domination is indeed universal, this fact is, from a political point of view, of paramount importance. The political achievement of feminism can be described, without exaggeration, as the discovery and definition of an entire political class ignored by more traditional political theories. Nevertheless, the political significance of patriarchy should not blind us to the fact that from a theoretical point of view the universality of male dominance is anomalous rather than central. A brief consideration of the reasons for this will help to forestall any tendency to see universal male domination as a motivation for a crudely biological theory of gender difference.[19]

Although the universality or near-universality of male dominance is a phenomenon well worth theoretical study, there is no reason for taking it as contradicting the basic variability in human sexually differentiated behavior that I have been asserting. In the first place, it is only one case to set against many. But this is a misleading way of putting the point; male domination is a phenomenon on a higher level of abstraction than is the characterization of particular forms of behavior in particular societies. There is no reason to suppose that the exercise of male domination is itself something that has always been implemented by the very same kinds of behavior. On the contrary, the kind of labor that women perform for men is quite different in, say, feudal societies, gatherer-hunter societies, and modern industrial societies. And the social institutions and personal interactions that enforce such performance are equally variable. Hence the implementation of male domination—which is what is crucial, rather than its mere existence, for assessing the plasticity of behavior—far from contradicting the variability of gender roles, graphically illustrates it. Analogously to my conclusion in the case of sex, there may be good reasons to suppose that human sex differences generate forces that have a strong tendency to result in male supremacy. But as in the previous case, although the presence of such a force may be of use in explaining the genesis of a particular gender-differentiated society, it does not identify any remotely homogeneous property that characterizes the present state of that society. In this case, even if male supremacy is genuinely universal, the enormous variability in the form it takes indicates extensive interactions with more specific forces that, in turn, show that there are no grounds for assuming that even the abstractly characterized consequence is in any way inevitable.[20]

Picking up now the main theme of the argument, as with the case of sex in general, we should now consider the question whether man and woman (in their biological sense) are potentially useful explanatory categories. As *purely* biological kinds, they are largely unexceptionable, if of very limited significance; it may be allowed as a modest nomic generalization, for example, that humans born with penises tend to grow facial hair later in their lives. But there is a very powerful tendency, which it has been my goal here to expose and criticize, to extend the relevance of explanatory categories beyond their empirically warranted limits, a tendency that derives philosophical nourishment from the idea that when one has distinguished a kind one has discovered an essence. If, in fact, the empirical significance of the kinds man and woman does not go beyond some systematic, if quite variable, physiological differences and the observation that men have achieved a dominant position in all or most societies, the kinds distinguished seem of modest significance. Nothing in those empirical facts provides motivation for thinking that those categories should be accorded fundamental importance in explaining the highly diverse forms of behavior found in very different social systems, whether such explanatory efforts are motivated by androcentric apologetics or by misplaced feminist enthusiasm.

The conclusion I am defending might be stated as follows. Just as the concept of sex in general will do little to explain why peacocks, but not turkeys, have long tails, or why the prairie chicken, but not the gander, is polygynous, so the concept of woman will do little to explain why Asian women once had their feet mutilated or why twentieth-century Western women are more likely to become nurses than doctors. In principle, the same move is open as was suggested at the end of the previous section. It would be possible to suggest that the explanatory categories should again be narrowed, so that for behavioral explanations the relevant classes would be as specific as female !Kung or upperclass Englishman.[21] At this point, however, the claim to have identified even the most attenuated strong natural kinds would be impossible to sustain. The identification of a strong natural kind must involve the belief that the behavior of its instances depends, in some important cases, on intrinsic properties of the individual characteristic of members of that kind. But it would be hard to find even the most ignorant racist nowadays prepared to assert that the dispositions to behavior and social interactions of a man raised in, say, rural Zimbabwe would have been just the same if that individual had been brought up in a wealthy Californian suburb. To concede that explanations must appeal to kinds with that degree of specificity is to concede beyond serious argument that it is local, presumably cultural, factors that are of primary importance in understanding behavior. There may be natural kinds here in the tolerant sense that I have advocated; but the point of that tolerance is precisely to leave it a wholly empirical question how broadly generalizations apply. Some may apply to such narrow groups, some to broader or even still narrower groups. Only experience can tell.

Categorial Empiricism

I will now elaborate, in the light of the preceding cases, the argument against essentialism sketched at the beginning of this chapter. The main force of the argument is a plea for complete empiricism with regard to the explanatory potential of particular kinds. My suggestion is that a commitment to real essences either is vacuous or violates this demand. In partial reaction to this point, I shall also suggest that less fundamental importance be attached to the delineation of kinds, and attention rather be directed toward the identification of properties, dispositions, and forces. To connect these points, what makes a kind explanatorily useful is that its instances share the same properties or dispositions and are susceptible to the same forces. But since we have no way of deciding how much such concomitance to expect in any particular kind, the discovery of a kind adds little, if anything, to the discovery of whatever correlations may turn out to characterize it. An essence can be seen as a promissory note on the existence of such correlations. It is a promissory note that empiricists should reject. I take the preceding discussion to illustrate this point in the following way: it is easy enough to distinguish classes at many different levels of generality—males, male vertebrates, men, Irishmen, and so on—but there is nothing in this process of differentiating classes that provides any basis for predicting the extent to which its members will be amenable to lawlike generalizations. Finally, this conclusion in no way impugns the theoretical significance of the *properties* on the basis of which such classes are differentiated.

The clearest example I have offered in support of this plea is the case of sex. What we find in this case is a major seam in nature that not only fails to distinguish robust natural kinds, but also fails to distinguish classes that realize any general lawlike regularities. The explanation of this seam is simple enough: it reflects a presumably uniform type of historical process rather than the discrimination of any causally uniform type of entity. But it is hard to see what could be the basis for postulating a natural kind, in the strong sense of a set of common possessors of a real essence, except either the perception of a natural seam among phenomena, or the discovery of one or more laws satisfied by a class of phenomena. In the first case, as the example of sex shows, the inference to a natural kind would be illegitimate; in the second case, unless it constituted a quite ungrounded assumption that further, hitherto-undiscovered laws were in the offing, it would be wholly redundant.

Having rejected the possibility that general sexual categories could provide a basis for lawlike generalization, I then considered the possibility that far more restricted sexually specific categories might still constitute natural kinds in the strong, essentialist sense. I do not claim that this possibility has been, or perhaps even could be, rigorously refuted. However, the issue of gender shows a rather different kind of failure of the assumption of essentialism. Here again, it is easy enough to see the reasons for this failure. Many factors affect human behavior, including behavior that is gender differentiated. For this reason it is foolish to expect that there will be forms of behavior that are determined simply by the agent's being, say, a human male, still less by the agent's being merely male. There is, moreover, no evidence even that *dispositions* to behavior are uniform among human males. Nevertheless, unless we are careful to restrict the import of our categories to the empirical, we are in danger of being led into just such assumptions. And indeed the danger in this case is not merely academic. There is a fair-sized industry of a priori speculation on the biological causes of human gender differentiation, an industry that is not only intellectually disreputable, but potentially even politically pernicious. (This claim will be amplified in the final chapter of the book.)

I do not take myself to have shown any particular limits on the nomological significance of sexually defined categories. In the case of humans, there are good reasons for doubting whether this significance extends much beyond the purely physiological. In many other species there are sound generalizations to be made about sexually dimorphic behavior. My point is not that sex is a scientifically useless concept, but rather that from a conceptual standpoint that seeks natural kinds and their underlying essences, one is very likely to misrepresent that significance. I should now, therefore, say something more about how I think that significance should be understood.

To begin with, nothing I have said contradicts the idea that sex is a highly important *property*. In other words, sex is a property that, in a sufficiently specific context, is frequently an indispensable part of lawlike generalizations. At a certain time of year, for instance, the males of a certain species of birds produce characteristic and predictable noises. If you know the time of year and the species of bird, what you additionally need to know, if you want to predict whether it will make that noise, is what sex it is.

Undoubtedly of more theoretical interest are the ways in which sex connects with evolution. At the most theoretical level, there is the question of the origin and maintenance of sexual reproduction. Since sex seems, prima facie, such an extraordinary waste of time from the point of view of the females,[22] this is a very baffling question. On the other hand, this very problematic nature of the phenomenon makes it likely that there is some powerful evolutionary process at work. At a less general level, as Darwin emphasized at great length, the existence of sex can have profound effects on the particular course of the evolution of a species. Thus, I am far from denying the biological interest of sex. What I want to claim is that the way in which the basic sexual categories—male, female, neuter, hermaphrodite—divide the natural world tells us nothing about either the extent to which such categories will give rise to general laws or, more importantly, what will be the range of application of whatever interesting laws do involve those categories. It is the denial of this latter point that, I believe, is required to provide any motivation for an essentialist position and that, I have argued, is very difficult to reconcile with the range of phenomena I have been discussing.

Let me make one further comment about natural kinds. There is certainly no harm in calling a set of objects that are found to have a substantial number of shared properties a natural kind. The discovery of such a kind, however, provides no basis for the supposition that some particular properties can nonarbitrarily be singled out as essential. But there is no reason why the term *natural kind* should be wedded to essentialism—or, anyway, no more reason than an accident of linguistic history that could readily be rectified. It is, after all, a natural term for distinguishing butterflies, coprolites, volcanoes, and suchlike somewhat homogeneous naturally occurring kinds of objects from kinds of objects that are either not naturally occurring (artifacts) or not remotely homogeneous (property of the U.S. government, fast food, bigger than my head, and so on). With this proviso, I am quite happy to refer to species as natural kinds. This case is unusual, in that we do have reason to believe that members of a species will share a large number of properties, this reason being that we suppose the members of a species to have come about through a uniform historical process. However, this concession in no way contradicts my insistence that the extent of homogeneity within even so homogeneous a kind should be treated wholly empirically. Members of a species also vary considerably. And we cannot know a priori how variable any particular feature will be.

The only way that we could provide grounds for dispensing with this empirical stance would be if we were somehow to know that the members of certain kinds were completely homogeneous in all respects (or in some set of respects somehow distinguishable a priori). Many people seem to believe this to be true of the kinds distinguished by physics and chemistry, although I find this doubtful.[23] If these doubts are unwarranted, physics and chemistry are, in an important respect, very different from biology. But even if this is the case, it is surely an empirical fact, not something that could be known a priori. It is entirely possible to conceive of a world composed of indivisible atoms, each as different from one another as one organism is from the next. Such would not appear to be our world; but *how* homogeneous physical or chemical particles may be remains an empirical matter. If this assertion is correct, even microphysics cannot provide a hiding place from the categorial empiricism I am advocating.

II. REDUCTIONISM

4. Reductionism and Materialism

Probably the most powerful source for the idea of science as a unified project is the commitment to substance monism: the belief that, in some important sense, the world is composed of some one kind of stuff. There are few contemporary idealists, so the specific version of substance monism to be considered here is materialism. The importance of contemporary materialism for my present theme is that it is often taken to necessitate some kind of reductionism. And reductionism is the traditional, as well as the most powerful, grounding for a belief in the unity of science. In this chapter I shall look at contemporary materialism, especially the version that is now more often referred to as physicalism, and investigate how various materialist theses are related to various versions of reductionism. I shall argue that only the most innocuous version of materialism is plausible, and that this kind of materialism provides no basis for any version of reductionism. In the remaining chapters in this part of the book, I shall examine some more specific reductionist theses, first in biology, and second in psychology and the philosophy of mind.

Introduction to Reductionism

Reductionism, in its broadest sense, is the commitment to any unifactorial explanation of a range of phenomena. Thus, for example, the (generally outmoded) variety of Marxism that holds that all social phenomena should be explained in terms of economics is, in this sense, a reductionist program. In this broad sense, it is generally assumed that the reductionistic is a subset of the simplistic. Here I mean something rather more specific, namely, the view that the ultimate scientific understanding of a range of phenomena is to be gained exclusively from looking at the constituents of those phenomena and their properties. Versions of such a view still have many adherents among philosophers and scientists.

The sense of reductionism with which I am concerned constitutes a family of views, the members of which I shall attempt to distinguish in the course of the next four chapters. To begin with, it will help to introduce what is perhaps both the clearest and the strongest version, which I shall call classical reductionism.[1] Assume, first, a hierarchical classification of objects in which the objects at each level are complex structures of the objects comprising the next-lower level. Paul Oppenheim and Hilary Putnam (1958) propose the following levels: elementary particles, atoms, molecules, living cells, multicellular organisms, and social groups. The investigation of each level is the task of a particular domain of science, which aims to discern the laws governing the behavior of the objects at that level. Reduction consists in deriving the laws at each higher (reduced) level from the laws governing the objects at the next-lower (reducing) level. Such reduction, in addition to knowledge of the laws at both the reducing and reduced levels, will also require so-called bridge principles (or bridge laws) identifying the kinds of objects at the reduced level with particular structures of the objects at the reducing level. Given the transitivity of such deductive derivation, the endpoint of this program will reveal the whole of science to have been derived from nothing but the laws of the lowest level and the bridge principles. The lowest level will be the physics of elementary particles. Thus, finally, truly basic science need concern itself only with the objects described by particle physics. To serve as a reminder that we do not yet know exactly what these are, or even that they are not quite misleadingly referred to as "particles," I shall refer to them henceforward, rather inelegantly, as "physical entities."[2]

As I shall discuss shortly, there are great difficulties with this account, and there are also a number of variations on this simple reductionist position, some designed to address some of these difficulties. But first, I want to consider the general grounds for adopting any reductionist thesis, of which I believe two are crucial. One of these is the supposedly successful elaboration of fragments of the reductionist program in the past history of science; the other is the prior commitment to a metaphysical thesis, materialism, that is taken in some way to entail reductionism.

In a limited sense, I do not wish to question the fact that reductionism has had great successes. Many of the greatest achievements of science depend essentially on insight into the structure of objects. But the significance of this fact is very unclear. It is debatable and debated whether such structural insights imply anything like derivation of the behavior of those objects whose structure is elucidated. Moreover, and more importantly, *whatever* is the nature of the illumination gleaned from knowledge of structures, it does not follow that *all* scientific inquiry has importantly to do with elucidation of structures. In fact, I shall argue that this is not the case. In subsequent chapters I shall attempt to throw some light on the question, What are the uses and limits of structural explanation? But for now I shall look more closely at the second main ground of reductionist assumptions, materialism.

Materialism

Materialism is the currently dominant version of metaphysical monism, the ancient view that underlying the superficial plurality and diversity of the world's contents, there is a more fundamental unity and uniformity. Whereas the first part of this book was intended to establish the existence of many overlapping but equally real classifications of the objects at one level of organization (the biological), here I want to argue for an equally liberal pluralism across structural levels of organization. Cats and dogs, mountains and molehills, I want to argue, exist in just as metaphysically robust a sense as do electrons and quarks (assuming, of course, that the latter exist at all). What I want to deny, therefore, is that there is any interesting sense in which ontological priority must be accorded to the allegedly homogeneous stuff out of which bigger things are made. Although there is an interpretation of materialism that I am willing to endorse, it is an interpretation that makes materialism a much less powerful doctrine than is generally supposed, and that provides no obstacle to pluralism of the kind just indicated.

There are two major obstacles to taking pluralism seriously as a metaphysical doctrine. The first is mainly historical, the legacy of Cartesian dualism. The influence of this metaphysical system has been so strong that most of the debate about materialism has posed dualism and monism as the only serious contenders. The most fundamental ontological issue is taken to be the question whether the mind is really just an arrangement of matter, or whether it is a distinct kind of substance. Given the general disrepute into which dualism has fallen among philosophers, it has seemed evident to many that monism is the only serious option. What is often overlooked is that a significant part of the intuitive implausibility of dualism is addressed as successfully by pluralism as by monism. It is strange and implausible to suppose that there should be just one domain of phenomena so anomalous as to require the positing of an irreducibly distinct substance; but if minds are no more nor less anomalous than cells, societies, or weather systems, the basis of this implausibility is largely undermined.

The second main source of antipluralism is the exaggerated deference that is sometimes accorded to the physical sciences. Thus it sometimes seems to be supposed that the physical sciences constitute, or at least give us good grounds to anticipate, the detailed elaboration of a monistic account of the world. Although the physical sciences have undeniably had some remarkable successes, such an anticipation seems to be based on a gross exaggeration of their achievements. The main response to this exaggerated assessment of the physical sciences must await the detailed demonstration, in this and the next three chapters, of the inadequacy of reductionism, which, in turn, will reveal clear limits to the scope of the physical sciences. In the meantime I shall aim to distinguish a plausible core of materialism from some of its more ambitious elaborations. This core is entirely compatible with ontological pluralism.

Here a terminological note is needed. So far the metaphysical view I am considering has been referred to only as materialism. Contemporary discussions, especially in the philosophy of mind, much more commonly discuss physicalism. This neologism is intended to emphasize a commitment to whatever are the objects of study of the science of physics, as opposed to some more traditional account of matter in terms of pure extension, impenetrable masses moving through the void, or whatever. On the whole I think this innovation is more confusing than illuminating. Physicalists universally admit that it is whatever account physicists finally converge on that will provide the ultimate criteria of existence, and refuse to be held hostage to the fortunes of any particular current physical theory. While earlier materialists were no doubt often more willing to anchor their metaphysics to particular theories, even taking aspects of such theories to be knowable a priori, the spirit of materialism has always been to remain sensitive to the best-motivated contemporary views of the nature of physical things.[3] Thus I do not see that our present rejection of seventeenth- or eighteenth-century accounts of matter should require contemporary physicalists to distance themselves significantly from the traditional spirit of materialism. I shall not, therefore, distinguish sharply between my uses of these terms. I shall, however, tend to use the term *physicalism* in reference to issues in the philosophy of mind and *materialism* for the more general thesis of which the former is a part. Although this distinction does not well capture historical nuances of usage, it may perhaps serve to remind us that matter has been conceived sometimes in opposition to mind, and sometimes in opposition to form.

The next task is to distinguish two very different ways of formulating a general materialism. The first, and less ambitious, might be expressed with the thought that if one removed from the universe all the physical entities (in the sense defined above), there would be nothing left.[4] This version accounts readily for the concern with the metaphysical status of the mental: minds are perceived as possible counterinstances to the monistic thesis; physicalist construals of the mental are intended to defuse this objection. Unfortunately, even if the proposed version of materialism is conceded, it is quite inadequate as an articulation of monism. The simple pluralist response is that it claims only that by taking away the physical entities one would, somehow or other, take away all the other kinds of things too, which hardly shows that those other things are nothing but the physical entities of which they are composed. In a similar way one might bring an end to a game of bridge by physically carrying off all the players; which hardly entails that the game is nothing but the four players who currently constitute it. The appeal of this kind of materialism stems from purely negative consequences. The nonexistence of mental substance, God, the afterlife, or whatever, I am happy to concede. But these denials of existence do not require any very strong positive thesis of monism. Provided the alternative is seen to be a pluralism based not on different kinds of stuff but on irreducibly different kinds of things, and provided we reject the false dilemma of materialism versus Cartesian dualism, this kind of materialism is irrelevant to the issue between monism and pluralism.

Can a more sophisticated, and perhaps more consequential, statement of this kind of weak materialist monism be found? A somewhat more promising statement would be that whatever kinds of things there may be, they are all made of physical entities. An immediate difficulty is that there are many kinds of things, such as political systems, the rules of chess, and irrational numbers, that do not appear to be made of anything at all. Perhaps there are good reasons for attaching ontological primacy to those things that are made of something, in which case the claim that whatever is made of anything is made of physical entities is both plausible and not entirely trivial. It is not trivial, because it again denies what Cartesian dualism, for instance, has affirmed, that minds are made of some other kind of stuff. And no doubt it is possible to believe that gods, numbers, abstract entities, or whatever, are made of something, and of something other than physical stuff; or perhaps even that every kind of thing is made of its own unique kind of stuff. However, this statement of weak materialism is no more help than the preceding one as a defense of monism against pluralism. In the first place, it gives us no reason to attach any kind of ontological primacy to those things that are made of anything; and second, and most importantly, it does not provide any reason for attaching preeminent metaphysical—or for that matter scientific—importance to what things are made of. Modulating from the contrast with mind to the contrast with form, why should we emphasize matter so strongly to the exclusion of form? I shall call this plausible, but relatively uninspiring, metaphysical thesis "compositional materialism."

The second version of materialism embodies a much bolder suggestion, and one that overtly or covertly underlies a great deal of philosophical work on physicalism: the proposal that everything that happens can be explained, at least in principle, in terms of physical entities and the laws that govern their behavior. This thesis does provide exactly the rationale for taking materialism seriously as a monistic metaphysics that compositional materialism lacks. The obvious prima facie objection to this thesis is that, insofar as they are explicable at all, biological phenomena must be explained in terms of biological laws or principles, economic processes in terms of economic laws, and so on. To meet this reply, what the materialist needs is a strong form of reductionism: the claim that all these various laws can be reduced to, or explained by, physical laws. Pending a detailed discussion of reductionism, for now I note merely that the required version of reductionism is, to say the least, controversial. Thus this way of converting the relatively uncontroversial (at least among philosophers) thesis of compositional materialism into something with more metaphysical bite also converts it into a vastly more problematic doctrine. To the extent that reductionism seems dubious, so should this kind of metaphysical monism. I shall refer to this thesis as "reductive materialism."

Corresponding to these versions of materialism it is possible to discern very different conceptions of ontology. A rather naive conception of ontology that might suggest some additional philosophical significance for compositional materialism is the idea that an ontological theory is just an inventory of the contents of space and time. In relation to this conception it can be seen that compositional materialism allows a major gain in conceptual economy for such an inventory. Suppose, for example, that a certain area of space contains a tree. The relevant part of one's ontological inventory might then read: one tree, one trunk, one hundred branches, ten thousand leaves, and so on. But insofar as everything on this list is composed of physical entities, there is perhaps a sense in which if we just listed all the physical entities, we would not have left anything out. Again we preserve the modest substantive claim allowed to compositional materialism, that there is nothing such as minds, God, and so on, that has been entirely overlooked by our list of physical entities. There is, of course, a great deal more to be said *about* what exists than can be found on this list, most importantly, facts such as that a certain set of the physical entities constitutes a leaf.

Whatever one thinks about this conception of ontology, if that is all there is to ontology, there is a great deal more than ontology to metaphysics. Its deficiencies parallel those mentioned for compositional materialism as an elaboration of monism. The conception of ontology that connects more closely with reductive materialism, on the other hand, is altogether more substantial. It is also influential in contemporary philosophy. The idea, particularly associated with Quine,[5] is that we are "ontologically committed" only to whatever entities we need to appeal to in order to explain anything. Prima facie, this conception will commit us to acknowledging all manner of diverse entities. However, if *reductive* materialism were true—that is, if we could explain everything by referring only to physical entities—this conception of ontology would provide a clear sense in which only physical entities need be admitted to exist.[6] An important corollary is that if this conception of ontology is accepted, and reductionism is shown to fail, we are immediately committed to a radical ontological pluralism.

In summary, merely compositional materialism, though important in making certain negative claims, does nothing to establish any interesting kind of monism. What is needed for a robust monism is precisely reductionism.[7]

Some Varieties of Reductionism

There are several rather different reductionist positions. The first distinction to note is that between synchronic and diachronic conceptions of reductionism.[8] The term *reduction* is often used to refer to the relation between a theory and its historical successor. For example, there has been considerable discussion of whether Newtonian mechanics can be reduced to a limiting case of relativistic mechanics. This is an example of what I mean by diachronic reductionism. My concern, on the other hand, is solely with synchronic reductionism, that is to say, with the relations between coexisting theories addressed to different levels of organization. This distinction is particularly important in relation to the discussion of reduction in genetics, to be taken up in Chapter 6. The relation between molecular genetics and Mendelian genetics is often treated as if the former were simply a historical successor of the latter.[9] But while it is no doubt true that *some* of the questions that were once considered part of Mendelian genetics are now taken to be within the subject area of molecular genetics, this historical progression is not my present topic. Rather, I am interested in the relations between the *current* theoretical approaches to the very different kinds of questions (for example, about molecular events in individual development, about the transmission of phenotypic traits, or about changes in the genetic endowments of populations) associated in some way with genetics. More specifically, I shall argue against the common reductionist assumption that all these questions will be given a unified treatment with the development of molecular genetics. It is natural for a reductionist to blur this distinction, since if one sees reduction as a central ingredient of scientific progress, it is natural to see it as the ideal fate for successful scientific theories. Thus reduction is simultaneously historical and contemporary. Those more skeptical of reductionism should bear the distinction in mind.

The second distinction is perhaps the most significant for general metaphysics and, most especially, the philosophy of mind. Although a large number, perhaps even a majority, of philosophers of science have given up on the practical possibility of reduction, many still want to insist on its remaining, in some sense, a theoretical possibility. Obstacles to reduction are seen as "merely practical," so that the possibility of reduction remains for sufficiently large computers or, at least, for the divine mind.[10] The distinction between theoretical and practical obstacles to reductionism, however, is by no means as simple as it might seem. The paradigmatic conception of a practical obstacle is mathematical or computational complexity. But, first, this appears to make the practical/theoretical distinction a hostage to the current level of development of our mathematical techniques; and, second, we may have no way of telling whether a mathematical problem is insoluble in principle or simply for us given our current mathematical and computational resources. Even this latter distinction is not wholly clear. For example, it is sometimes suggested that a problem might be computationally tractable, but only by a computing device with more components than there are elementary particles in the universe. This would seem to be a contingent problem (the universe might have been bigger), but it is not naturally described as "merely practical." Finally, the situation has been further complicated by recent attention to chaotic systems, systems that are deterministic but infinitely sensitive to their exact initial conditions.[11] It is certainly possible that chaotic behavior at a lower level might thwart reduction, although it is again unclear whether this should be seen as a practical or a theoretical obstacle.

The tendency to give up on practical reductionism but to maintain theoretical reductionism is most evident in the current prevalence of supervenience theses. The proposal that one domain of phenomena (D1) supervenes on another (D2) is intended to capture the idea that although there are no systematic links between the two domains, the former depends entirely on the latter. Thus it is said that given the state of D2, the state of D1 is also given (though, importantly, not vice versa), or that there can be no change in D1 without some change in D2 (and again, not vice versa). In the case of reduction, of course, D2 will be the lower level potentially reducing D1.

Supervenience entered philosophical discourse as an account of the relation of moral properties to natural properties proposed by philosophers such as G. E. Moore and R. M. Hare.[12] That these are the philosophers most closely connected with the denial of the "naturalistic fallacy," the attempt to define moral concepts in terms of nonmoral properties, makes it clear that they conceived such supervenience as nonreductive. More recently supervenience has been discussed extensively as a possible formulation of nonreductive physicalism in the philosophy of mind.[13] More recently still, it has been proposed as an account of the relation between levels of organization in biology, especially between organismic and population-level phenomena and genetics.[14]

Since supervenience has widely seemed attractive precisely as a way of avoiding reductionism, it will be apparent that my characterization of it as a weak version of reductionism is controversial. The sense in which that supervenience is a kind of reductionism can be explained as follows. If a level of organization H supervenes on a lower level L, and if God knew the complete state of things at level L, then she could infer the complete description of the state of things at level H. This is just the kind of thing I mean by reductionism in principle.[15] I assume that the reason I am inclined to see this thesis somewhat differently from many philosophers is that they find the general metaphysics of reduction so compelling as to present supervenience as the weakest imaginable relation between higher and lower levels of organization. Since I advocate a more egalitarian conception of the relations between levels of organization, I see supervenience merely as a very weak, but still questionable, form of reductionism.

The claim that the relations between higher and lower levels of organization must be at least as strong (that is, as reductionistic) as supervenience must depend on the intuition that supervenience, at least, is self-evidently true. The only question left open is whether some stronger, more forthrightly reductionist thesis might be true as well. (Supervenience is, of course, entailed by more ambitious reductionist theses.) But supervenience is not *self-evidently* true. It is surely imaginable, for instance, that people with identical physical states, including states of the brain, might be thinking different things. This possibility is defended by those who believe that the content of a thought typically depends on facts external to the thinker.[16] *Evidence* for supervenience, it seems, would have to be the kind of evidence necessary for reductionism. It would be evidence that higher-level phenomena are indeed determined by lower-level phenomena, or that identical (or sufficiently similar) lower-level phenomena do indeed produce the same higher-level phenomena. As is the case with evidence for reductionism generally, the problem is that where such evidence exists at all, it is in a narrow range of quite specialized cases, and the legitimacy of extrapolation to a general philosophical thesis is, to say the least, questionable.

The third and last major distinction between kinds of reductionism is a distinction within the category of practical reductions. Even adherents of the classical reductionism described above are likely to admit that in the process of carrying out a reduction it is likely that the reduced theory will undergo some changes and so, strictly speaking, is not being deduced from the reducing theory. If the changes to the reduced theory in the process of reduction are sufficiently extensive, it may become more appropriate to say that the higher-level theory has been shown to be false and been replaced by an extended version of the reducing theory.[17] Thus we can identify two extreme cases: at one extreme the classical reductionism described above, which could be called reduction by derivation; and at the other extreme reduction by replacement or elimination. In fact neither of these extremes is very plausible. A reduction by derivation is likely to involve some amendments to the reduced theory, and a deduction by replacement is hardly conceivable without some fairly successful theory of the higher-level phenomena to provide a starting point.[18] Thus these extremes are best seen as defining a spectrum of cases involving more or less alteration to the antecedent theories at the higher level. Philosophers will immediately recognize some major positions in recent philosophy of mind paralleled in this distinction. Whereas contemporary physicalist views of the mind began with claims by U. T. Place (1956) and J. J. C. Smart (1959) clearly in the spirit of derivational reductionism, replacement or eliminativist suggestions followed fairly quickly (Feyerabend, 1963; Rorty, 1965, 1970) and have recently become the dominant view (P. M. Churchland, 1981, 1985; Stich, 1983; P. S. Churchland, 1986).

The existence of this range of positions makes reductionist claims somewhat slippery to oppose, for any reason for supposing that a certain domain of theory cannot be derived from the appropriate lower level is liable to encounter the response: "So much the worse for that theory; it will have to be replaced by something less intransigent." This indeed is essentially what occurred during the development of physicalist theories of the mental. Arguing directly against replacement theories is difficult for the well-known reasons that make negative existential claims hard to establish. The claim that a higher-level theory will eventually be eliminated in favor of an extension of a lower-level theory, an extension that in every currently extant case is admitted not yet to exist, is precisely an unproved existential hypothesis.

The situation, however, is perhaps not really as difficult as this exposition suggests. The strongest arguments against derivational reductions are likely to involve some account of why nothing that could be said in terms of lower-level entities could even in principle capture some aspect of the higher-level phenomena. If this kind of argument can be established it will, of course, apply equally against the possibility of replacement. Most of the more specific antireductionist arguments in the following chapters will be addressed, in the first instance, to reductionism conceived as synchronic, practical, and derivational. It will be important, however, to bear in mind the question whether the arguments also demonstrate the infeasibility of "in principle" reductionism and replacement reductionism.

An Argument for Reductionism

The argument that must be addressed here turns on the requirement that explanations at different structural levels must at least be consistent with one another. Certain views about the nature of causality suggest that only some kind of reductive relation between higher and lower levels can achieve such consistency. Although it is not easy to find direct statements of such an argument, I believe that it underlies a good deal of continued insistence on at least the weaker versions of reductionism.[19] It also provides a link between two of the pillars of the mechanistic picture of the world that it is the aim of this book to oppose. Although I am here considering how an assumption about causality entails reductionism, it can equally well serve as an argument from the falsity of reductionism to the falsity of the view of causality in question.

The crucial assumption about causality is what I shall refer to as the assumption of causal completeness. This is the assumption that for every event there is a complete causal story to account for its occurrence. Obviously enough, this is a view of causality the roots of which are to be found in the soil of determinism. The paradigm of a complete causal story is the sufficient (and perhaps even necessary) antecedent condition provided by a deterministic

causal explanation. However, since the issue of causal completeness at the microphysical level is central to the present discussion,
and since microphysics is generally agreed, in the light of contemporary interpretations of quantum mechanics, to be irreducibly
indeterministic, it is important to consider the indeterministic analogue of deterministic causal completeness. It is not hard to see
what this should be. The basic idea is that there should be some
set of antecedent conditions that together determine some precise
probability of the event in question.[20]

Events and their causal explanations are conceived at particular
levels of organization. The mutual annihilation of an electron and
a positron is an event at the subatomic level; my thinking of the
previous example is an event at the mental level; a supernova is
an event at the astronomic level; and so on. However, events at
lower levels may very well be constituents of events at higher
levels, and it is this that leads readily from the commitment to
causal completeness to an argument for some kind of reductionism. The argument is most easily seen in terms of a simple example. Consider an electron in my index finger. As I move my
finger to type the letter *b* on my keyboard, the electron must
move. Causal completeness at the level of elementary particles
implies that there is some condition of events at that level sufficient
to explain the movement of the electron. Suppose, then, that I
offer some causal story at the macroscopic level about the movement of my hand. For example, suppose my hand moved because
I intended to type the letter *b,* or, for those who deny that this is
a causal explanation, suppose that my finger was dragged down
by my typing instructor. Either story, if it explains the movement
of my finger, also entails the movement of the electron. At the
microlevel, meanwhile, there are presumably sufficient causes operating on every elementary particle in my finger. The causal
efficacy of events at the macrolevel had better at least be *consistent*
with all these billions of microlevel causal facts. Such consistency
would appear to require either divinely preordained harmony, or
reducibility at least in principle. This point can be made clearer
by noting that the causal completeness of the microphysical directly contradicts the supposition that my intention (or coercion)
could be *necessary* for the movement of the electron, except insofar

as it is itself a necessary consequence of events at the microlevel. For the events at the microlevel are, *ex hypothesi,* themselves *sufficient* for the movement of the electron, or, in the indeterministic case, sufficient to determine the precise probability of its movement. So it appears that events at the macrolevel, except insofar as they are understood as aggregates of events at the microlevel— that is, as reducible to the microlevel at least in principle—are causally inert. This, of course, is the classical picture of Laplacean determinism, except that it does not depend on determinism, only the causal completeness, or causal closure, of the microlevel. And metaphysically, though perhaps not epistemologically, this picture entails the causal impotence of events at the macrolevel.

A connection should be noted between this argument and compositional materialism. The negative claims of compositional materialism are surely necessary conditions for the credibility of causal completeness at the microlevel. Leaving aside the possibility of epiphenomenalism (and with even fewer qualms, psychophysical parallelism), the existence of nonphysical minds with any causal efficacy in the physical world makes causal completeness at the microlevel impossible. This is simply because the appeal to reducibility in principle is precluded when the higher-level causal agent is not, even in a merely compositional sense, identifiable with any set of microphysical entities. Similar consequences follow from interventionist deities, ghosts, and so on. Without reducibility in principle, the same goes for interventionist people, animals, or even rocks.

My reaction to this should be clear from what has gone before. Although I want no truck with the nonphysical substances just mentioned, a central purpose of the ontological pluralism I have been defending is to imply that there are genuinely causal entities at many different levels of organization. And this is enough to show that causal completeness at one particular level is wholly incredible. By contrast with even the weakest versions of reductionism, the pluralism I have in mind precludes the privileging of any particular level. My response to the argument just described has two main parts. The next two chapters will explain in some detail why reductionism fails in major parts of biology and psychology. This can be seen, among other things, as preparing the inversion of the reductionist *modus ponens* (causal completeness requires reductionism) into my antireductionist *modus tollens* (the failure of reductionism implies the falsity of causal completeness). In Part III, I shall criticize the idea of causal completeness directly.

What Is Wrong with Reductionism?

Finally I want to introduce in a general way what I take to be the problems with global reductionism, and also to say a little about the possible virtues of more local reductionistic strategies.

A first difficulty, already mentioned, is the oversimplification involved in the hierarchy of levels of organization. To begin with, a very striking exclusion is the whole of macrophysics. Atoms or molecules go to make up not only living cells but also many complex nonliving things. At the very least, science must be depicted as a branched tree to accommodate gravitation, electromagnetism, cosmology, geology, and so on, although it is in fact not clear how even these branches should be rooted in microphysics. Nor can the objects with which the macroscopic life sciences are concerned generally be treated as structures solely of objects from the next-lower level. A sponge, perhaps, is a multicellular organism that *is* nothing but an assembly of cells. But an aardvark would more naturally be treated as an assembly of organs and other complex systems. Moreover, many of its essential constituents, such as hormones, are not cells, or parts of cells, but molecules, whereas ions, also vital for its well-being, are more nearly atoms. So we find not only an intermediate level of organization, but also items from lower levels. Nor can these lower-level entities be treated merely as exogenous environmental variables. Their presence may often be determined by very complex events at the level of organs or organ systems. Consider, for example, the increase of blood adrenalin of an animal in danger, a chemical consequence of perhaps a complex perceptual event. This illustrates what I shall emphasize in the next chapter, the possibility that lower-level events may perfectly well be determined by what is happening at a higher level.

To take a rather different case, there exist intricate ecological systems of bacteria. An ecological system, if it fits into the hierarchy at all, would appear to be a social group. But the largest constituents in this case being single cells, the putative reduction will have to miss out a level. Ecology is a particularly interesting case. Not only do ecological systems typically combine elements from at least three levels—multicellular organisms, single cells, and molecules (as nutrients in the environment)—but the understanding of such systems typically involves such factors as climate and geology, factors that would have to be assigned to parallel branches. I suspect that the complex interdependencies of entities at many different levels of structural complexity characteristic of biology is sufficient to show the implausibility of the reductionistic project. However, I hope to show that the explanatory privilege of the small and simple can be rejected even if these complexities are ignored.

Whereas the preceding points begin to illustrate the extent to which reductionism is grounded in an oversimplified view of natural phenomena, there is a deeper difficulty with the theory of derivational reductionism, a difficulty that is grounded in the conclusions about natural kinds and essentialism defended in Part I. I have mentioned that derivational reductionism requires bridge principles identifying types of structurally complex objects with particular structures of simpler objects. My major objection to this kind of reductionism is that the discovery of such bridge principles cannot be expected.

I have suggested that we call the kinds of objects discriminated by a science the natural kinds of that science. Recalling my proposal that the term *natural kind* be divorced from traditional connections with essentialism, I use this expression here to emphasize the following fact: a system of classification is typically an inextricable part of the science to which it applies.[21] One outstanding example is provided by the periodic table of the elements, which is at the same time the basis of an extremely powerful classification of chemical substances and the embodiment of a central part of chemical theory. Equally, the theory of tectonic plates determines a particular classification of geological events, and psychoanalytic theory implies a particular classification of psychopathologies. Thus to whatever extent we take our scientific theories to be "correct," the taxonomy they presuppose may be said to be "natural." This is, of course, a very weak interpretation of natural kinds. Apart from its lack of commitment to essentialism, the degree of naturalness of kinds will depend on the extent to which one thinks of scientific theory as independent of human interests, a question that might be answered quite differently from one area of science to the next. If any scientific inquiry is conceived of as addressing questions the significance of which is partly determined by particular human interests or needs, then the best possible theory in this domain will delineate only natural kinds relative to the interests in question. This is not antirealism. Nature may well determine what is the best theoretical approach to these problems. But other problems would imply other theories and other, equally natural, kinds. On the other hand, the characterization of natural kinds that I have given does not exclude the possibility of much stronger interpretations.

The strongest relevant interpretation of natural kinds here is one that assumes that there is some correct system of classification independent of any theory of the behavior of the types distinguished.[22] This is consistent with the foregoing remarks about the relation between classification and theory, since if there were such an independently correct classification, presumably any adequate theory would have to be framed in terms of it. Such an interpretation makes classification logically (if not necessarily epistemically) prior to any theory. The relevance of all this to bridge principles, finally, is just that the requirement that scientific terms be amenable to occurrence in bridge principles, a requirement on any higher-level theory if it is to be reducible, though not determining any particular taxonomy, provides a very strong constraint on what could be an adequate taxonomy. My general point is that such a constraint is intolerably strong.

This constraint implies at least that a bridge principle will provide a necessary and sufficient condition for membership of a kind in the science to be reduced. If the condition were not sufficient, it would be impossible to infer, from the perspective of the reducing science, that objects of the higher-level type would obey particular laws, since the reducing science would provide no way of determining that an object was of the relevant type. Equally, if the condition were not necessary we would be unable to infer the *general* conformity of objects of a particular type to laws of the reduced science, since the inference would extend only to those objects that happened to satisfy the condition. Bridge principles, therefore, must be at least as strong as biconditionals. Moreover, it is obvious that not just any biconditionals will do. What is needed is a necessary and sufficient condition on kind membership specified in structural terms. Reductionism, it appears, presupposes essentialism, and an essentialism in which essential properties are invariably structural.

I have already argued at length that many scientific kinds in the first place lack *any* essential property, or any necessary and sufficient condition on membership and, second, that the actual criteria for membership of many scientific kinds have nothing to do with physical structure. In an important sense, this observation is sufficient for the refutation of reductionism. However, there are several reasons why it is necessary to go into much greater detail about particular cases, which will be the task of the next two chapters. First, the argument to this point is extremely abstract. Apart from tending to make it proportionately unconvincing, it does not greatly illuminate what the rejection of reductionism involves in any particular case, or what insight it may be taken to offer about the nature of scientific inquiry at higher organizational levels. Second, and more important, the argument so far applies only to strict derivational reductionism. But I have already admitted that this is only one end of a continuum of reductionist positions, and that the most plausible positions lie nearer to the center of that continuum. What the more detailed discussions will aim to show is that the kinds of factors appropriate for determining classifications at higher levels in many cases have nothing to do with the structural properties of objects. Such nonstructural accounts of the grounds of appropriate classification will show the impossibility also of replacement types of reductionism. This demonstration will finally establish the genuine autonomy of higher-level sciences from the theories of their structural parts.

I must end with an important qualification. It is not my project to claim that explanation in terms of structure has no role to play in science. On the contrary, it is clear that many of the most prized of scientific achievements involve just such, at least loosely reductive, explanations. The problem, if my antireductionist arguments are correct, is to decide what are the appropriate roles for reductive explanations, and what are the cases in which they fail. A broad suggestion that, I hope, will be made more plausible in the sequel is the following: reductive explanation is required to account for *how* things of a certain kind do what they do; but they typically do not help us to understand or to predict *what*, among the behaviors of which it is capable, a complex thing will do. The latter is generally to be addressed in terms of the autonomous understanding of the phenomena at the higher level.

This suggestion can readily be illustrated with respect to the sciences of the mental. That neuroscience is relevant in important respects to understanding how the mind does whatever it does is not seriously disputed. But how to characterize correctly what the mind does is extremely controversial, and indeed continues to constitute the majority of work in the philosophy of mind. It is even controversial *what* it is that does whatever the mind does, with positions ranging from the neuro-Cartesianism of contemporary materialists, in which the brain takes over sole responsibility for the exact range of functions attributed by Descartes to the mind, to the more radical rejection of Cartesianism exemplified by Wittgenstein's aphorism that "the human body is the best picture of the human soul."[23] Whatever the answers to such questions may be, only some parts of the story will call for structural explanations of the kind to be hoped for from neuroscience. Much philosophical work remains to be done in deciding which parts might require such explanations. I shall offer a small contribution to that work in Chapter 7.

5. Reductionism in Biology 1: Ecology

The question of reductionism has been a dominant concern throughout the recent history of the philosophy of biology.[1] Commitment to some kind of reductionism sometimes seems to be taken as a precondition for biology to take its place among the genuinely respectable sciences. The issue is enormously complicated by the fact that biology is a discipline (or perhaps collection of disciplines) in which a considerable number of levels of organization are regularly distinguished and considered. These include, at least, genes, cells, organs, multicellular organisms, and groups of organisms ranging from family groups and demes, through populations and species, to ecosystems and higher classificatory taxa. Since genes are generally assumed to belong simultaneously to the domain of chemistry, and cells to that of biology, it is worth mentioning also such things as viruses or viroids, which hover uneasily on that boundary. Given this array of structural levels, there are a correspondingly large number of possible reductionist projects.

Reductionist thinking in biology has been unified, to some degree, by the rise of molecular genetics. Apart from more obvious achievements such as elucidating the mechanism of inheritance and some aspects of the processes of development, some biologists see genetics as providing the key to a unified understanding of evolution. Apart from the extreme version of this in the genic selectionism of Richard Dawkins (1976, 1982), it is quite widely supposed that population genetics, the study of changes in the genetic constitution of populations, is the area of inquiry that should illuminate the basic processes of evolution. In the next chapter I shall consider these issues in some detail and criticize what I take to be exaggerated estimates of the relevance of genetics to evolutionary theory. But first I want to clarify some general ideas about the possibilities for reduction by looking at an area of biology much less worked over by philosophers, ecology.

Ecological Models

By ecology I do not, of course, mean a generally favorable attitude to whales, clean air, and so on, but rather the study of the determinants of the abundance, or relative abundance, of particular kinds of organisms. Since biologists working in this area will, in fact, use standard taxonomic classifications to identify the organisms in which they are interested, the conclusion established in Chapter 2, that standard biological taxa have no defining features, should be sufficient to establish that ecology is not reducible by derivation from any level of science lower than biology. Since this will reasonably enough prompt the response that too strict a conception of reduction is being invoked, a further object of this discussion will be to show that the only appropriate principles of classification for this area of science involve appeals to properties at the macrolevel, so that the possibility of reduction by partial or total replacement is also undercut. Perhaps more important, I hope to bring out some general features of the relation between macroscopic processes and the microscopic events that constitute and, in certain respects, explain them. This will suggest a general diagnosis of what is wrong with reductionism, which will be supported in the discussion of genetics in Chapter 6.

Apart from a lot of purely descriptive work on the way organisms interact with one another and their environment, the theoretical core of ecology is the construction of mathematical models, based on empirical observation and plausible biological assumption, intended to reflect the development over time of interacting populations.[2] I shall illustrate this methodology with a simple but elegant example.

Certain wasps reproduce by injecting an egg into a live insect of some other species; the latter is then consumed by the developing wasp larva. Since insect populations can generally be treated as progressing in discrete generations, we can use difference equations to construct a simple model of the dynamics of an isolated system of such wasps and their hosts. We assume that every member of the host species either is parasitized or lays n eggs. Let the number of parasites in each generation be P_t, and of hosts be H_t. Assuming that every egg will hatch, and that every parasitized insect will produce a parasite[3] in the next generation, it follows that

$$(1)\ H_t = P_{t+1} + H_{t+1}/n.$$

Also, $H_{t+1} = nHt$ times the proportion of hosts that are not parasitized. If parasites search independently and randomly for hosts, we can assume that this will take the form

$$(2)\ H_{t+1} = nH_t e^{-aP}t.$$

Combining (1) and (2) gives

$$(3)\ P_{t+1} = H_t(1 - e^{-aP}t).$$

The model thus gives the two populations in each generation as a function of the populations in the preceding generation.[4] With plausible values of the parameters n and a, the model shows the populations oscillating with ever-increasing amplitude to extinction. And in fact if such a system is set up in the laboratory, this is said to be what typically happens. The number of parasites increases until a generation is reached in which every host is parasitized, when, of course, both host and parasite species die out. Needless to say, additional conditions need to be incorporated into the model if it is to be applied to the real world in which such interacting populations manage to survive. For example, it might be that as host numbers decline they become concentrated in areas where they are increasingly difficult to find, thus implying a more sophisticated density dependence for the parasites' search function.

Before considering such models further, it will be useful to mention a distinction that has been proposed between abstract and general systems of laws.[5] Paradigmatic examples of general systems are Newtonian mechanics or Maxwell's electrodynamics, while abstract systems are said to be exemplified by electrical network theory or functional psychology. The supposed importance of the distinction is that whereas general laws specify the kinds of objects to which they apply, and thus genuinely explain the behavior of those objects, abstract laws specify only the behavior of systems of idealized objects that fulfill certain functional conditions implicit in the laws. Thus it is claimed that such laws are not fully explanatory, since we also need reasons for supposing that the objects concerned satisfy the specified functional conditions, or at least approximately do so.

Like the distinction between derivational and replacement reductions discussed in the previous chapter, this distinction should be seen as introducing a continuously variable property of laws rather than as dividing them sharply into two disjoint categories. The reason is simply that all systems of laws require for their strict truth classes of objects, homogeneous or precisely specifiable in all relevant respects, to which they apply; whereas, in reality, no class of objects is truly homogeneous in all such respects. The continuum from abstract to general laws reflects the degree to which it is supposed that an identifiable class of objects can be precisely specified with respect to the properties relevant for the application of the laws. We are inclined to think that the relevant properties can be fully specified for massive objects subject to Newtonian forces, so that Newtonian mechanics is a natural paradigm for a general system of laws. However, influential arguments by Cartwright (1983, chap. 3) suggest that even this assumption is simplistic. Real objects will typically be subject to various causes of motion, for example, so that laws such as the law of gravitation will be true only of idealized sets of objects to which none of these further factors applies.

At any rate, whether or not there are systems of laws that fully realize the conditions for being general systems, it is clear that the population model just outlined lies well to the abstract end of the continuum. We need now to ask, therefore, to what extent it is legitimate to apply such a model to real ecological systems. Of course, the model was not derived as a random stringing together of symbols; it was constructed to reflect the actual interactions and behavior observed in certain kinds of organisms. On the other hand the model clearly embodies an idealization of such observations. The host insects, for example, are no doubt liable to other kinds of demise, and certainly not every insect lays exactly *n* eggs. Such idealizations, though showing that there are limits to the fit between the model and the objects to which it is applied, certainly do not show that it might not be a respectable and useful bit of scientific theory. It might be useful, for example, in developing sophisticated approaches to pest control.

However, it would be quite unrealistic to expect the methodology exemplified by such a model to lead to indefinitely greater perfection of explanatory and predictive power or to convergence on a truly "general" system of laws. A crucial explanation of its limitations is in terms of the conflict between the taxonomic principles implied on the one hand by the abstract structure involved, and on the other by the practical demands of the concrete application. In the latter case we have actual, countable populations of particular kinds of organisms, and in the former, theoretically defined terms such as *parasite, prey,* and *competitor.* It is hardly likely that these could ever be wholly extensionally equivalent.

Perhaps the most famous of such ecological models concerns a cyclical variation in the populations of Canadian lynxes and hares.[6] Without going into details,[7] let us suppose that the best current model for this system gives reasonably reliable predictions about these populations, at least assuming that major extrinsic variables (weather, epidemics, and so on) stay within normal bounds. Any vaguely realist account of the success of this model will need to identify prey in the abstract model with hares in the concrete realization; and the observation that lynxes sometimes eat rabbits, say, will suffice to show that this identification is less than perfect. A little imaginative elaboration of this seemingly banal observation will help to illustrate the more general and perhaps equally banal point, that there are many, probably vastly many and variously variable, features of lynxes, hares, and rabbits that are relevant to their success in, and means of, surviving and reproducing, but not considered in the abstract ecological model.

An imaginable strategy for rectifying this situation would be to employ a taxonomy based more directly on the theoretical content of the terms in the model. That is, instead of trying to estimate the number of hares, we might go about identifying lynx prey. Although this strategy might be acceptable in some cases, in others it might defeat both practical and theoretical goals of the exercise. The distinction between rabbits and hares is, indeed, one that from most biological perspectives, including phylogeny, is trivial to the point of invisibility. It is, nevertheless, one that is commonly drawn by experts neither technically scientific nor scientifically technical, such as farmers, hunters, and amateur naturalists. And the phylogenetic triviality of the distinction by no means excludes its importance to (some of) us (as discussed at length in Chapter 1) or, more important still to the present argument, its having some central theoretical importance in some other context. Suppose, for instance, that our concern with the whole question derives from an interest in the deleterious effects of burrowing on the stability of the landscape. At the heart of the distinction, such as it is, between hares and rabbits is the fact that the latter but not the former are disposed to excavate large communal residences. Suppose, then, that the prevalence of rabbit warrens was the determinant factor in the adaptedness to the terrain of some large ruminant extremely liable to break legs when running across heavily burrowed terrain. Or suppose that the solitary, vagrant life of the hare necessitated slight differences in the color or texture of the coat, which greatly reduced the value of the fur. In either case, information about the prevalence of lynx prey (hares and rabbits) without distinction between hares and rabbits might be completely useless for those purposes for which it was most urgently needed. The more general direction in which this leads is the epistemological correlate of the metaphysical pluralism recommended in Part I: one reason that scientific narratives constructed for different purposes will be incommensurable is that they will need to be told in terms of noncoincident kinds.

The immediate moral of the preceding story is that our interest in a kind of organism need not be restricted to those features that predominantly determine its abundance; and kinds of organisms that are highly similar with respect to the features and interactions that determine their abundance may nevertheless differ in respects that are of crucial interest to us. Thus, any attempt to develop models of ecological systems in terms of ecological natural kinds runs the risk of delivering information quite unrelated to the prevalence of organisms of the kinds in which we are interested. This difficulty suggests the question whether there is some classification of organisms ideal for the study of ecological phenomena. A natural response would be that we aim to identify kinds of organisms in such a way as to maximize the information conveyed by knowing that an organism belongs to that kind. Such a strategy, presumably, would maximize the probability that our investigation would provide information in terms of the kinds of organisms we are interested in knowing about and, hence, with the abundance of which we are most likely to be concerned.

This proposal is likely to engender various kinds of skepticism. First, it may be doubted whether the goal of maximizing the information content of a classification even makes sense. It might seem, for example, to raise the kinds of difficulties with counting properties that have seemed so telling against objective accounts of similarity and, thereby, against phenetic conceptions of taxonomy. Although I have no intention of defending an account of objective similarity, I think this difficulty is easily exaggerated. As the discussion that brought us to this point suggests, the information that we are concerned with is information that *we* (a particular, historically situated group of humans) might possibly care about. The possible grounds for such interest are, indeed, extremely variable. They might include properties that are economically useful or strikingly noticeable, and they certainly might include properties that are of interest for further theoretical reasons. But though variable, the condition of human interest or theoretical significance surely provides a very substantial limitation, or at least a pragmatic ranking, on the range of properties of special relevance to taxonomy. The adjustment of classificatory practices in response to such more or less significant distinguishing properties, the significance of which derives from a diverse and pluralistic set of interests, cannot but be a central aspect of the evolution of taxonomy.

The second objection to the conclusion that ecology will best be served by a general taxonomy conceived in conformity with the discussion of the preceding paragraph comes from the currently dominant assumption that taxonomy should reflect phylogeny. I have already expressed my reservations about this assumption and will not repeat them here. Of course, phylogeny will tend to generate classifications close to those that capture maximum similarity of the members of kinds or maximum information-carrying content of assignment to a kind. And since much biological inquiry is explicitly directed at reconstruction of, or elucidation of the processes within, phylogeny, phylogeny will very probably be the central determinant of theoretical weighting of potentially classificatory features. Nevertheless, as the rabbits and hares and many earlier examples illustrate, this is not the only relevant factor. To the extent that it is not so, there is likely to be significant conflict between the demands put on taxonomy by the various goals driving modeling projects such as those we have been considering. First, we have the inherent aims of the ecological inquiry; second, we have the pursuit of these goals in terms of biological kinds from a generally phylogenetically conceived taxonomy; and third, we have whatever further reasons there may be for interest in the ecological results. No taxonomy can ideally mediate these conflicts, and this impossibility is perhaps the clearest way of seeing the ineliminable imperfectibility of such models. A phylogenetic taxonomy, since it introduces a classificatory goal largely external to the ecological project, is likely to be even further than necessary from the ideal.

The Failure of Reductionism in Ecology

A closer look at the relation between the individual events that contribute to the determination of population numbers and models in population ecology will serve to develop an argument that, quite apart from the irreducibility of laws concerning the members of biological kinds to laws of lower level, laws concerning the dynamics of populations could not be derived from any amount of information about the properties of the individuals that constitute those populations.

Let us assume first that there is some reliable criterion by which we can identify lynxes; call it P. (From the point of view of this discussion we may even assume that P is an essential property of lynxes.) A crucial property of lynxes we would need in order to derive the macromodel for the dynamics of the lynx population from the properties of its members would be the propensity of lynxes to eat hares. Since not every lynx eats every hare, this must be a propensity that is exercised under restricted conditions. For example, the lynx must be sufficiently hungry, or perhaps its kittens are, the hare is sufficiently close, and the lynx is aware of it. It is also obvious that even if there were some precise such set of conditions that determined of a particular lynx that it would catch a particular hare, we would not expect the same conditions to apply to every lynx: some lynxes do not have hungry kittens, some can run faster than others and therefore catch more distant observed hares, and so on. Moreover, some hares, such as elderly sick ones, will no doubt be easier to catch than others, who might be young and healthy. And finally it is more than doubtful whether even a complete specification of all the relevant properties and capacities of both animals would suffice to determine whether the hare gets eaten: one or the other might trip, the hare might run into a cul-de-sac or discover an escape hole, and so on.

All this is quite irrelevant to the success of the macromodel, which requires only that the mean propensity of an arbitrary lynx to eat hares is reasonably determinate or, more accurately, a determinable function of such things as the population levels of the animals. But this is certainly not a property uniformly associated with P; there is no reason why any lynx should have just this propensity. *A fortiori,* no physiological explanation of why lynxes eat hares will provide an explanation of the fact that a particular lynx has this particular property. The relevant property, in fact, is strictly a property of the population or species, and cannot contribute to a reductive explanation of the laws involved. The difficulty may be seen as constituting a dilemma. The reduction would need an attribution to lynxes of a determinate propensity to eat hares. This could be interpreted as a statistical property, or as a property of individual lynxes. But the former line leaves us still with a property at the level we are trying to reduce, whereas the latter is simply false of most of the animals to which it is attributed. I now want to suggest a more general diagnosis of where reductionism has failed. This will also raise a further pressing question to which I shall propose the outlines of an answer: if reductionism is false, what account should be given of the evident illumination that has been derived from microexplanations?

To put it at its simplest, the general failure of reductionism may be attributed to the following fact: the individuals that would have to be assumed for the derivation of the macrotheory cannot be identified with those that are the subjects of descriptive accounts at the next-lower level, although their relationship may be close enough to allow such derivations to serve important explanatory purposes. The possibility of this nonidentity is to be explained by the fact that the individuals at both levels are idealizations. Both models at the macrolevel and descriptive accounts or laws at the microlevel involve abstractions. But the abstractions involved are not the same. In relation to my argument in Part I, this point can be seen as establishing the possibility of autonomy for higher-level natural kinds.

The examples discussed above demonstrate that the idealized individuals that form the basis of models in population ecology are distinct from those that might be described by physiologists or behavioral psychologists. These latter sciences also aim to describe idealized individuals. Physiologists will attempt to abstract from idiosyncrasies of the individuals that they judge to be irrelevant to the typical structure of organisms of that kind. Such abstraction might aim at different degrees of generalization; thus, a physiologist might aim to describe the liver in a way that applied either to a large group of organisms that have livers or merely to the (typical) liver of a particular species.[8]

Two important points need to be made about the relation between the properties that are relevant to forming our (idealized) accounts of individuals, and those that are relevant to the microtheory underlying a higher-level theory such as population ecology. Most obviously, the difference in our interests will focus our attention on properties of quite different kinds. In the former case our interest may be aroused directly by properties or behaviors of the organisms that attract our attention or excite our curiosity. In the latter case, on the other hand, the properties of concern are determined by the macrotheory we are trying to construct. If we are interested in population numbers we are forced to observe, and make hypotheses about, the properties that affect the survival and reproduction of members of the species. What makes this distinction especially significant is that there is no reason to assume that all properties are equally amenable to systematic study at the level of the individual: it is not only properties that vary within a species, but also the degree of variability of different properties. It is typically the observation that a certain property is quite *consistently* displayed in a certain group of organisms, at least under specifiable circumstances, that leads us to investigate the structural or other basis of this property in the individuals. On the other hand, when our interest in a property is extrinsically determined by its relevance to a macrotheory, we have no reason to assume that this property will be a stable feature of the members of the kind in question, and no reason, more generally, to expect that property to be amenable to systematic investigation at the individual level. Just as the possibility of bridge principles provides a strong constraint on the (reducible) macrotheory, we now see that the macrotheory extrinsically constrains what could be an adequate (reducing) microtheory.

The second point is related, though more subtle. This concerns the relation between behavioral capacities and actual behavior. Consider again the observed fact that lynxes eat hares. In explaining the *capacity* of lynxes to enjoy such a diet, much can be said at the level of the individual. They must have suitable teeth and digestive systems; appropriate perceptual, cognitive, and athletic abilities; and so on. But although these capacities may indeed be amenable to structural explanation, they are certainly not sufficient to determine the outcome of a particular interaction between a hare and a lynx. I have already made the obvious point that this must at least depend on the environment and on the hare. But in fact it strikes me as quite implausible that we could ever construct a theory that predicted the outcome of each such encounter, or even that attributed a determinate probability distribution of outcomes to such encounters.[9] Certainly the possibility of such a theory is not presupposed by the practice of population ecology. But in its absence, we are left with nothing but the average propensity of lynxes to eat hares, a property that emerges only at the macrolevel. This point provides a very general diagnosis of the limits of reductionism. We should expect the behavioral capacities of an entity to be amenable to some kind of mechanical, structural, or broadly reductive explanation. The exercise of those capacities is another matter, and will typically depend on factors at the same structural level as the entity in question, or perhaps more likely, simultaneously at a variety of structural levels.

The preceding remarks also help to dispose of a weaker position to which a reductionist might hope to retreat. Such a reductionist might accept that the indefinite variability of phenomena forced us to make do with somewhat blurred classifications that did not entirely coincide with the homogeneous classes implied by our abstract systems of laws. But she or he might still wish to reinstate reductionism as a relation between the idealized individuals described by science at the various structural levels. On this picture, science would attempt to provide a unified core, to which reality would approximate to the extent that actual kinds coincided with ideal, abstract classifications.

My objection to this picture is simply that even the ideal kinds do not coincide between levels. The ideal hare that the physiologist might construct out of ideal cells is just not the same as the ideal hare that is hunted by the ecologist's ideal lynx. Indeed, as should become much clearer in the discussion of genetics in the next chapter, the core concepts of science are exposed to conflicting pressure not only from different structural levels but also from different kinds of inquiry at the same level. And so the unity of science cannot be maintained even for such an ideal core.

This conclusion leads back to a question raised a few pages earlier: In the absence of reduction, what is the importance of microexplanation? A short answer in the present case is, not very much. This issue will be pursued in more depth in the discussion of genetics, where structural explanation has played a much larger and less questionable role. However, the basic outlines of what I take to be the answer can be seen even in the case of ecology. Knowledge of the constituents of a complex process may not so much enable us to derive the laws that govern the process, as either tell us what kind of process it is or tell us how it is possible at all. Reductionistic explanations typically tell us what something is capable of rather than what it actually does. Thus, if lynxes did not have suitably sharp teeth they would be unable to indulge a taste for hares. A second reason for the importance of microexplanation (or, as we might often more perspicuously characterize it, microdescription) is that an understanding of the underlying structure may, though it need not, lead to improvements in the macrotheory. Both these effects of microexplanation should become clearer in the discussion of genetics; for the time being, they can be illustrated with a bit of science fiction.

Suppose that we have observed for some time a distant planet that displays a cyclical color change from green to brown. We might also have invented a mathematical formula that predicts what proportion of the planet will be green at any time as a function of the proportion that was green at some past time. Later, with better telescopes, we discover what is actually going on. The planet is populated by green algae and algivores. The population of algivores turns out to depend on the amount of algae in such a way as to generate oscillations of population numbers. Although this new understanding might do nothing to improve our ability to predict the course of these phenomena, it would surely tell us what we did not previously know, what this phenomenon really was and how it was possible. This case differs from more familiar ecological cases only in its simplicity and in the artificial obscurity of the nature of the process. Nevertheless, the parallel is clear. Fluctuations in population densities, especially of economically significant species, are often observed. Prima facie, these could perfectly well be, for example, random fluctuations. The kinds of model I have been considering could be seen as, *inter alia,* an argument for an understanding of such fluctuations in terms of an underlying ecological structure.

That insight into ecological phenomena can be gained by a greater knowledge of the microlevel is less than completely obvious, if only because the "micro" phenomena are often discovered in rather simple ways, such as by standing patiently and attentively behind a tree. And so in this case the macrotheory is generally constructed from an antecedent knowledge of much, at least, of what goes on at the microlevel. But it is surely plausible that we can always hope to find out more about the actual interactions that occur in a particular system, and that this information is likely to lead to greater sophistication in our ecological models. We can easily imagine, for example, that distinguishing different varieties of algae on the imaginary planet might lead us to more-sophisticated and more-accurate representations of the ecological events there. On the other hand, it is equally possible that such further knowledge would do little to enhance our understanding of the macroscopic phenomena, or that the limits of such enhancement would soon be reached.

6. Reductionism in Biology 2: Genetics

In a merely rhetorical sense, ecology as an example suffers from the defect that it is not a part of science particularly likely to inspire reductionist enthusiasm. On the other hand, if any recent scientific developments have tended to make reductionism look good, they are the spectacular advances that have been made in molecular genetics. The elucidation of the chemical structure of DNA and the subsequent elaboration, sometimes in exquisite detail, of some of the chemical processes involved in both reproduction and development are rightly counted among the major achievements of twentieth-century science. Insofar as this constitutes the explanation of a set of macroscopic phenomena by means of an account of the microscopic events of which they are composed, it would appear to provide the materials for a paradigm of physical reduction. There has been a fair amount of philosophical debate about the extent to which this case does bear out a reductionist thesis. Surprisingly, perhaps, the preponderance of recent opinion seems to have favored a negative conclusion.[1]

A natural point of entry to this debate is to ask, What exactly is it that is to be reduced to what? This question arises because a number of different scientific domains are immediately related to genetics, and molecular genetics suggests the possibility of a number of distinct reductionist projects. First, there is the reduction to molecular genetics of classical Mendelian or transmission genetics, encompassing the investigation of phenotypic patterns of inheritance, and its explanation in terms of hypothetical genes arranged on chromosomes. This has been the dominant concern of philosophical discussions of reductionism in genetics. Second, there is the explanation of the processes of ontogeny, the way in which the development of the phenotype is guided by the "program" embodied in the DNA. This is the project for which a broadly reductionist approach is most uncontroversially appropriate, a fact that perhaps explains the much smaller amount of philosophical attention accorded to it. And third, there is the reduction of population genetics, the field of inquiry that is generally taken to describe the basic mechanisms of evolutionary change. This reductionist project relates in obvious ways to the idea of treating the entire evolutionary process as fundamentally involving the differential survival of genes, in the manner advocated by Dawkins (1976).

The existence of these various possible reductionist projects suggests a parallel with the thesis of the previous chapter. We see in the case of the gene, as appeared for individual organisms in the preceding chapter, that the most interesting biological concepts and kinds of biological entities occupy nodes in our conceptual scheme where various projects of inquiry intersect. I argued for the case of organisms that, in contrast to the convergence of these projects on antecedent natural kinds, which would most readily further the ideals of reductionism, such nodes are much more complex than they might at first appear. The classifications best suited to the goals of these various projects will often diverge significantly, so that instead of a preexisting natural kind we find a number of partially overlapping kinds, and a taxonomy that compromises as best it can between a set of competing demands. A similar situation applies to the various investigations clustering around genetics. Expressed in the terms developed in the previous chapter for the case of ecology, the grounds of my antireductionist position are as follows: the genes described structurally by the molecular geneticist are not the same things as those referred to in the models of population genetics or even of classical transmission genetics.[2]

The central focus of the philosophical debate over the reductive potential of molecular genetics has been the relation between the referents of the term *gene* in Mendelian and molecular genetics. And the predominance of antireductionist opinion referred to above has been grounded on the view that the relations between the Mendelian and molecular senses of *gene* are at any rate too complex to permit reduction.

Since all the scientific investigations I am here considering require distinguishing between genes of various kinds, a precondition of any such reductionist projects is some classification of stretches of DNA.[3] Since DNA is—though in a way this is the seemingly innocent first step that leads to all the trouble—a chemical, it is natural to assume that the fundamental classification of DNA must be in terms of precise chemical formulas. (Chemistry, after all, provides our most plausible candidates for truly essential properties.) So we imagine that ideally every kind of gene should be defined as a certain sequence of base pairs.[4] It is undeniable that the chemical conception of genes has been indispensable in investigating such fundamental biological phenomena as replication, mutation, and the expression of segments of DNA as polypeptide chains. Moreover, it appears that the assumption of transmission geneticists before the discovery of DNA was that genes were specific segments of chromosomes.[5] Since chromosomes are now known to be strands of DNA, the conception of genes under consideration therefore fits exactly with the intentions of classical genetics.

Nevertheless a number of philosophers have refused to identify the genes referred to by molecular descriptions of stretches of DNA with the genes referred to in transmission genetics. Various grounds can be distinguished for this refusal.[6] The most clearly understood point of this kind, made very forcefully in early critiques of reductionism by David Hull (1974), is sometimes referred to as the "many-many problem." This problem arises with any attempt to identify genes characterized in molecular terms with genes distinguished by their relation to the phenotype. The problem is simply that, however we divide up an organism into phenotypic traits (itself a problem without any easy solution), it appears that many genes will typically contribute to the production of a trait. Moreover, if we consider any (molecular) gene, there will normally be a range of phenotypic traits to the development and character of which it contributes. (These phenomena are referred to, respectively, as polygeny and pleiotropy.) But it is in terms of genes identified in relation to phenotypic characters (wrinkled peas, red-eyed *Drosophila,* and so on) that classical transmission genetics has developed. Key concepts, such as dominance and recessiveness, concern the conditions for the appearance of phenotypic characters. Given, then, the many-many relations between molecular genes and phenotypic characters, the reduction of transmission genetics promises to be at least prohibitively complex.

Kenneth Waters (1990) argues that the force of this point has been exaggerated. As Waters notes, the developers of Mendelian genetics were well aware of the complexity of the relationship between genes and traits. Thus the complexity of the relationship between molecular genes and phenotypes is a prerequisite for, rather than an objection to, any adequate reductive explanation of the complex relations between Mendelian genes and phenotypes. The fallacy in the objection is that it is posed as if the Mendelian gene for, say, red eyes was the (sufficient) causal basis for red eyes. But in fact, as Waters notes, Mendelian genes were never intended to explain the production of traits, but only differences in traits between members of a particular population (that is, against a particular genetic background). "The gene for *X*" is simply a way of referring to some gene, in the sense of a stretch of DNA, the presence of which increases the likelihood of the presence of *X*. Thus Richard Dawkins writes: "When a geneticist speaks of a gene 'for' red eyes in *Drosophila* . . . [h]e is implicitly saying: there is variation in eye colour in the population; other things being equal, a fly with this gene is more likely to have red eyes than [is] a fly without the gene" (1982, p. 21).

I am not sure to what extent antireductionists have really been misled by the failure to keep in mind this understanding of the Mendelian gene. (I shall suggest below that it in fact does present serious difficulties for reductionists.) At any rate, it is clear that further analysis is required of what bearing the many-many relations between genes (Mendelian or molecular) and traits has on the various questions about reductionism in genetics. I shall approach this question from a somewhat different angle, by considering the relevance of a dichotomy rather different from our identification of kinds of genes, that between structure and function.[7]

Recent philosophical analyses of biological function have tied functions to an appropriate evolutionary etiology.[8] Schematically, the basic idea is that F is the function of a trait T if ancestral organisms with T were selected because having T enabled them to F, which in turn contributed to their reproductive success. This account connects our intuitions that, for example, the function of wings is to permit flight or of lungs to exchange gases with the air, with our beliefs about the evolutionary origin of such traits. For the purposes of the present discussion, however, I do not need to insist on such an account. The only aspect requiring emphasis is that whereas the structure of an entity looks downward, as it were, to its constituents and their relations, the function of an entity looks upward to its role in a system or entity of which it is part. Thus although no one would now suggest that it was the function of a lynx to devour hares, the contrast between the physiological (structural) account of the lynx and the account of the lynx required for an ecological investigation is analogous to the contrast between the function and structure of an entity such as a gene.[9] In the present context, the only significance of referring to functions is that to have a function an entity must play a role in some more complex system. Of course, something that satisfies this simple condition will frequently satisfy the more stringent definition of function as well. As we consider various investigations concerned with the functions of genes—or perhaps merely with their roles in more complex biological systems—the adequacy of a purely molecular classification of genes is increasingly jeopardized.

Even with gene expression considered at the level of polypeptide chains, the chemical characterization of the gene will normally be too narrow to be useful. First, there is redundancy in the genetic code, so that various sequences of base pairs can give rise to the same amino acid. Thus in an ideal taxonomy substitution of synonymous codons should not affect the classification of a DNA sequence; all that matters is that a sequence is among those that code for a particular amino acid. This complication indicates the need for a merely irritatingly disjunctive elaboration of the chemical scheme. But it is only the beginning. Consider, for example, human hemoglobin. Since our concern with hemoglobin is as a constituent of an organism, we are interested not in the production of a specific complex polypeptide chain, but rather in a functional biochemical species; and the latter is not a chemical kind at all, but a fairly complicated class of chemicals without sharp boundaries.[10] Even when we have inscribed the canonical genetic basis of human DNA in the data files of the genome project, the exact polypeptide sequence for which that chemical gene would code will not be a necessary condition for being human hemoglobin. Variation in the precise structure of a protein is common, especially when it has no significant effect on the protein's function. It would be absurd to suppose that minor deviations from some arbitrarily selected norm disqualify a molecule from belonging to the kind of which its behavior is characteristic.

One might then hope that the set of possible innocuous mutations would at least be finite and well defined, so that the disjunction of possible chemically precise "genes for hemoglobin" still has some claim to existence in principle. Unfortunately, we next need to note that substitution *salva potentate* is as problematic as its more notorious linguistic relatives. There may be many forms of hemoglobin that are functionally equivalent, but there are certainly some that are less efficient. In the present case there is the familiar example of sickle-cell anemia. Is sickled hemoglobin not hemoglobin at all? But even to deny, somewhat implausibly, that seriously defective hemoglobin is really hemoglobin does not help. It is merely a constraint on where we are prepared to draw the line; it does not obviate the necessity of drawing an ultimately arbitrary line. For surely there may be many slightly less efficient variants of hemoglobin unknown, because irrelevant, to medicine, which, if known, could not possibly be denied the status of hemoglobin. It seems, then, that in the sense of *hemoglobin* relevant to biology, *hemoglobin* must refer to a variable and somewhat arbitrarily restricted set of chemical compounds.

So far I have considered DNA solely in relation to the simplest possible role in ontogeny, the production of a functional polypeptide chain. It will be clear that somewhat different considerations arise when we move from this to the transmission of phenotypic traits. First, although the many-many problem may apply to some degree even to the production of polypeptide chains (see Hull, 1974, pp. 40–41), it will become vastly more pressing at the gross phenotypic level. Compounding this difficulty with the need to define traits in functional terms at the organismic level can be seen, to put it mildly, to present formidable problems for the reductionist program. Of particular interest are some of the key concepts of transmission genetics: dominance and recessiveness, linkage, independent assortment, and so on. Kitcher (1984c) directs the main thrust of his antireductionist argument at such concepts by arguing that the level at which the phenomena of gene replication are appropriately explained is in terms of gross features of chromosomes rather than the "gory details" of the underlying molecular events.

Kitcher develops this argument by claiming that meiosis, the process in which the genetic material is reduced by half to form the haploid gamete, should be seen as one instance of a broader natural kind, a PS (pair separation) process. Any process of this kind, instantiated in the diploid eukaryote cell by the separation of its paired chromosomes, is capable of explaining perfectly such phenomena as linkage (on the same chromosome) and independent assortment (on different chromosomes). The fact that in meiosis this process is realized by members of a certain class of molecular events is irrelevant to the force of such explanations. Waters (1990) objects that as soon as we focus on the details of the biology it is no longer plausible that "the shallower explanations of [classical genetics] are preferable to the deeper accounts provided by molecular theory" (p. 131). In particular, Kitcher's pessimism is unwarranted when we note that advances in molecular genetics have in fact illuminated details of, for example, the mechanism of crossing over responsible for recombination and partial linkage.

Since I am suspicious of any suggestion that broader generalizations are inherently preferable to narrower ones, I have some sympathy with Waters' reservations. However, I think that Kitcher's argument can be recast in terms that mesh nicely with the antireductionist perspective I have been trying to develop. It is interesting, first, that a parallel exists between Kitcher's argument and well-known functionalist arguments against reductionism in the philosophy of mind.[11] Functionalists have insisted that a particular mental state is the state it is if it has a particular role in the overall mental economy of its possessor, and regardless of whether it is realized in a neural net, an array of silicon chips, a system of cogs and pulleys, or Cartesian mind-stuff. Thus the reduction of the mental to the physical fails to capture the full scope of the psychological generalizations it purports to explain. I would like to emphasize a similar broadly functional interpretation of the chromosomes in Kitcher's argument.

Chromosomes provide one more excellent example of the Janus-faced character of scientific terms that, I want to claim, is as characteristic of genetics as it is of ecology. Molecular genetics will undoubtedly be the source of insight into the structure and capacities of chromosomes. When, on the other hand, we consider broader systems in which chromosomes play a part (or have a function), this kind of illumination will be of only limited significance. What is needed to clarify the force of Kitcher's argument is a more specific statement of the contexts in which meiosis and the like are being considered. It is possible to read Kitcher's argument as if he were claiming that molecular investigations were irrelevant to any aspect of our understanding of meiosis. This, indeed, seems to be the way Waters reads it, and he is surely correct in objecting to any such claim. Rather, the issue is that the mechanism of meiosis is irrelevant to many of the contexts in which we are interested in the process. We should think instead, therefore, of a context in which it is significant what chromosomes *do*. Consider, for instance, a general theoretical discussion of sexual reproduction and the cost of meiosis. The point is not merely that the gory details of molecular events are irrelevant in such a context. It is rather that different kinds of inquiry will require different classifications of chromosomes or chromosomal events. When Kitcher's PS processes are considered from the perspective of their function in a larger system, they form, for certain theoretical purposes, a natural kind. From the perspective of structure, which is indeed the right perspective from which to address questions about how a particular PS process or a particular structurally homogeneous kind of PS process works, PS processes as a whole form a largely miscellaneous collocation. As is generally the case, classification cannot be divorced from the specific purposes for which it is intended.

Such an analysis becomes even more compelling when we

move to the broadest context in which genetic investigations have commonly taken place, and consider the genetic basis of evolution. First, it is undisputed that evolution selects functions rather than structures; thus, wherever there are functionally homogeneous but structurally diverse gene products, the classifications of genes relevant for evolutionary investigations will be structurally heterogeneous. Moreover, the appropriate principles of genetic classification for evolutionary inquiries may well differ from those best suited to the analysis of the role of DNA in the development or transmission of phenotypic traits. In the latter cases we are concerned with the way in which particular biochemicals contribute to the production of an organism or the way in which traits are passed on to progeny; whereas in the former case we are interested also in comparing the functioning of organisms with different traits or biochemical resources. These projects need not give rise to the same principles of classification; and certainly neither needs to provide principles of classification that generate chemically homogeneous kinds.[12]

To make clear why ontogeny and phylogeny will in fact motivate different classifications of genes, it is necessary to focus once again on the relation between genes and relatively complex phenotypic traits.[13] This is the point in the argument at which the many-many problem really begins to bite. Most actual reference to genes, at least in evolutionary contexts, is in terms of "genes for" various morphological or behavioral features. At the extreme is the talk in sociobiology of genes for such things as aggression or even upward mobility.[14] More sober-sounding entities, such as genes for red eyes in *Drosophila* or genes for albinism, are central components of the geneticist's stock in trade. As I have mentioned, geneticists do not suppose that such expressions identify the complete causal antecedent of a trait. Rather, they take those expressions to refer to some gene, present but not universally so in the population, the presence of which raises the probability of the occurrence of the trait. (In genetically relatively homogeneous populations it may raise the probability from 0 to close to 1.) This conception is, in certain respects, well suited to Dawkins' project of understanding evolution in genetic terms. First, the definition is set up so as to make it likely that if there is selection for the phenotypic trait, there will be selection for the genetic trait.[15] Second, the definition suggests no problem in principle in moving from an apparently straightforward case such as red eyes in *Drosophila,* to genes for aggression, or for playing Defect in prisoner's dilemma. Whatever the virtues and problems of this method of identifying genes, it is ultimately extremely uncongenial to a genuinely reductionist perspective on evolutionary phenomena.

One of the main consequences of Dawkins' definition is to distance its user from any suggestion that a piece of DNA is *the* causal basis of a phenotypic trait.[16] Presumably any of a large number of genes causally involved in the determination of eye color could meet the condition of raising the probability that a fly will have a particular eye color. The distance of the proposal from the naive idea that a gene provides the causal basis of a trait can be emphasized by noting that there are no genes for traits that are constant in the population. This curious fact also shows the extent to which the definition is framed with evolutionary dynamics in mind: it is those genes that may be subject to selection pressures that concern us.

That this definition of a gene for a particular trait is an embarrassment to the reductionist is clear if we contrast it with the cases most encouraging for genetic reductionism. In the sorts of cases in which genetics seems to contribute to a real understanding of the details of biological phenomena, such as the etiology of sickle-cell anemia or particular well-studied mutations in *Drosophila,* we really do have particular genes providing structural explanations for the occurrence of particular traits. On the other hand, when we start talking about genes for extremely complex phenotypic traits such as intelligence or aggressiveness or, for that matter, physical size or disposition to heart failure, there are large, perhaps vast, numbers of DNA segments that satisfy the definition for genes for these traits. The class, say, of genes for intelligence must encompass not only the more or less minor structural differences just discussed in connection with hemoglobin, but also segments of DNA from many quite distinct parts of the genome. This might well amount to a substantial proportion of the genome. The subset of these genes that would satisfy the official definition of "gene for intelligence" is the presumably very large subset of these that exhibit some variation in the population. Talk of such genes is unrelated to any attempt to illuminate structural processes that ground the development of intelligence. And the division between those that do, and those that do not, vary in the population is absolutely irrelevant to such a reductionist project.

Clearly, different projects of inquiry relating to genetics can imply quite different ways of classifying the genetic material.[17] And once again, the kind of scientific pluralism that this discussion is intended to motivate depends in no way on an antirealist perspective on the referents of the classificatory terms relevant to these various projects.[18] This situation, to the extent that it is typical of scientific inquiry, is sufficient to establish the intransitivity of reductive explanation. The kind of classificatory abstraction relevant to a structural explanation of the chemical capacities of (in this case) DNA is quite distinct from that required for an account of the role of DNA in the functioning of organisms or the course of evolution. Just as I claimed in the previous chapter for the physiological and ecological perspectives on organisms, this lack of fit between the structural and functional perspectives on the gene will provide an impassable obstacle to reductionist explanation. Although there are certainly important reductionist projects developing in the application of molecular genetics to the understanding of ontogeny and of the mechanisms of inheritance, there are adjacent scientific projects that involve nothing like reductionism. Reductionism is a local condition of scientific research, not an irresistible tide sweeping the whole of science into an increasingly orderly pattern.

Population Genetics

The preceding argument has aimed to show that classical transmission genetics is irreducible to molecular genetics. The elucidation of the genetic basis of ontogeny, on the other hand, would seem to be as clearly legitimate a reductionistic project as one could hope for, although even here, to the extent that one aims to describe phenotypic endpoints of development, one will be forced to consider classes of genes defined in terms that may make little sense from a strictly molecular perspective. Whatever the limitations of reductionism in these areas, however, it would be absurd to deny that insight into molecular genetics has greatly increased our understanding of these realms of phenomena. In this section I want to examine population genetics, and I shall argue that here even such a qualified endorsement of reductionist methods is dubious. Moreover, I want to suggest that the very coherence of the genetically grounded account of evolution implicit in population genetics is called into question by doubts about reductionism.

Since neither genes nor individuals themselves undergo evolutionary change, it is natural to assume that evolution involves changes in species or other lineages of organisms. Moreover, it is commonly claimed that evolution consists in changes in frequencies of various kinds of genes within a lineage, typically a population. But since it is also generally agreed that the basic process of evolution, natural selection, acts directly not on genes but on whole organisms—selection of genes, that is to say, is generally a consequence of selection of organisms—this view of evolution is rendered problematic by doubts about reductionism. The conception of evolution as consisting in changes in gene frequencies is the basis of the supposition that the part of science concerned with the fundamental processes of evolution is population genetics. I shall argue, on the contrary, that population genetics is, in important respects, an embodiment of reductionist mythology.

One aspect of population genetics, the investigation of the distribution of genes in various populations, is purely empirical, and has been important to theoretical issues such as the extent of genetic variation and the prevalence of heterozygosity. Even more important, such studies have played a central role in undermining any pretense at providing a scientific basis for racism. What I want to address here, however, is what is more generally taken to be the central part of population genetics: the construction of models, in terms of changes in the frequencies of genes, of evolutionary processes.

Population genetics aims to describe the trajectories of frequencies of particular kinds of genes in biological populations. It is reasonable enough (though not beyond question) to consider change in gene frequencies to be a necessary condition of evolutionary change. But whether evolution involves any interesting regularities in these terms is another matter. Population genetics, as is characteristic of model-building methodologies, starts with the simplest cases and works toward greater complexity. Thus the simplest population genetics models concern single genetic loci undergoing selection. Of course, nobody supposes that such models are structurally similar to typical processes in nature. But the project of developing more complex and hence more realistic models from this starting point must assume something nontrivial about the relation of such simple models to reality. A natural assumption in this context is that the simplest possible real systems are indeed structurally similar to these simple models. And this seems plausible: one can perhaps appeal to the ubiquitous sickle-cell anemia or to other standard cases that recur in textbooks and philosophical discussions. But such an assumption is far from sufficient. For this concession to have any bearing on more complex cases, one must also believe that the more intricate dynamics governing most natural genetic frequencies are, in some sense, built up from processes of the simpler kind. And this, of course, is a classically reductionist assumption.

In view of my preceding claims about the intransitivity of reduction, the reductionism inherent in population genetics should seem particularly suspicious. This is because the levels involved, genes and populations, are clearly nonadjacent. Populations are composed of individual organisms, whereas genes are some of the components of individuals. The coherence of population genetics requires that the abstract individual organism implied by (possible) accounts of genetically regulated development is identical with the individual that provides the canonical member of a given population. And in accordance with the general antireductionist maneuver that I have been advocating, this identity is unlikely to obtain.

To give a sense of the peculiar nature of population genetics, it might help to consider the parallel but obviously bogus imaginary science of molecular mechanics. By this I do not mean the mechanics of molecules, but rather the standard mechanics of macroscopic objects, but understood in terms of the determinants of motions of the molecules of which those objects are composed.[19] We can (sometimes) explain the motions of macroscopic objects in terms of the motions of the parts from which they are constituted as mechanical systems. And there are properties of those mechanical parts that might be explained in terms of properties of their constituent molecules. But the conceptions of those parts relevant to those two projects are distinct, and there is no reason to suppose that they can serve to connect these projects so as to provide a molecular explanation of the mechanical behavior of the whole. The parallel reinforces the suspicion that evolution (like mechanics) is a science that needs to be conceived in terms of macroscopic processes. A more detailed examination of the methods of population genetics will serve to substantiate this suspicion.

Consider how models of change in gene frequency at even a single locus under selection are constructed. Such models treat selection as consisting of the different probabilities that genes of different kinds will be transmitted from one generation to another. These probabilities are expressed with parameters known as selection coefficients, or just fitnesses, values of which parameters are taken to reflect the effect of possession of that particular gene on the survival and reproduction chances of organisms that possess them. But if the models that employ such parameters are to reflect the real structures of natural processes, then these coefficients must represent genuine properties or dispositions of the objects to which they are apparently referred, genes. Without offering a detailed theory of the reality of properties, in the present context we can surely demand that the presence of a particular gene will license an inference to a level of fitness characteristic, other things being equal, of that gene. And this is a demand that gene fitnesses are quite unable to satisfy.[20]

To substantiate this claim requires some careful attention to what is included in Dawkins' *ceteris paribus* clause above (page 124). It is generally understood that population genetics models must assume a constant external environment, and I shall allow this assumption for the present. What we cannot reasonably be asked to hold constant is the genetic environment. It is perhaps true that if we had a population uniformly homozygous at every locus except one, we could assign stable fitnesses to the alleles at the variable locus. And experimental investigations of highly inbred laboratory populations may even approximate this condition. But our present concern is with natural populations, and these will normally have many variable loci. Attempting to hold fixed the genetic background would buy us stable selection coefficients at the intolerable cost of invoking vanishingly small classes of organisms over which they were constant. A classic investigation of the grasshopper *Moraba scurra* by R. C. Lewontin and M. J. D. White (1960) revealed strong interactions between the measured fitnesses of two chromosome inversions. If such interactions are remotely typical of organisms with substantial genetic variability, individual genetic fitnesses must be seen, like the propensity of hares to get eaten, as no more than artifacts derived from averages over the many different real propensities in the population. And indeed, Lewontin concludes his discussion of these results with the remark that "We cannot [build] a theory of a complex system by the addition or aggregation of simple ones" (1974, p. 281).

It should hardly be surprising that genetic fitnesses display this strongly interactive character. Organisms, after all, are highly integrated systems. At the crudest phenotypic level it is obvious, for example, that the fitness benefits to an organism of having sharp claws or teeth might depend on whether it possessed internal organs capable of digesting flesh. We should surely expect such interconnectedness of function to go all the way down to the fine balance of cellular chemistry and, *a fortiori,* to apply to any phenotypic consequences of a particular variable gene. This argument would apply even if organisms exhibited perfect "beanbag" genetics, that is, if each gene produced exactly one phenotypic effect regardless of genetic context. Given, finally, that in reality genes interact in complex ways in the production of phenotypes, the radical dependence of selection coefficients on the genetic context can hardly be exaggerated. The consequence of all this is that even if population genetics models were to show significant empirical success in modeling changes in genetic frequencies in a population, they could not be interpreted realistically. There are few entities in nature possessing the stable selection coefficients of the entities apparently referred to in population genetics models. So it is improbable that bits of DNA could be such entities.

One last point should make this conclusion still more compelling. I conceded at the outset that population genetics models should be relativized to a constant environment. But it should be noted that this is an extremely restrictive assumption. In reality, evolution by natural selection is both an effect and a cause of a changing environment, with genetic fitnesses changing constantly in response to the environment. Moreover, organisms themselves have profound effects on their local environments. According to one evolutionary theorist, "The . . . relativity of genes to environments is so fundamental that it implies the existence of two routes to fitness instead of one. Active organisms may bequeath either 'better (or worse) genes' for 'anticipated environments' or they may bequeath 'better (or worse) environments' for 'anticipated genes'" (Odling-Smee, 1988, p. 74). F. J. Odling-Smee concludes from this thesis that "it is misleading to measure fitness exclusively in genes. Instead we need a more comprehensive currency, one capable of combining both an organism's genetic and ecological contributions to descent" (p. 116). And, of course, the ecological contributions will themselves affect the genetic fitnesses. No doubt proponents of genetic fitnesses would be able to suggest further complexities for population genetics models to account for the dynamic relations between environment and evolving genome. However, these considerations seem to me sufficient to undermine any realistic interpretation of genetic fitnesses even with constant environments. Instead of pursuing this line of argument any further, I want to consider a quite different and more illuminating approach to understanding evolutionary phenomena. The superiority of this approach will also help to illustrate the meaning and implications of ontological pluralism across structural levels.

Without generally seeing this as a threat to the status of population genetics, a number of biological theorists have always insisted that selection is a process that occurs, in the large majority of cases, at the level of individuals.[21] Selection, it is said, "sees" only individuals. It is individuals that are selected (reproduce, die) as a whole, and it is individuals, unlike genes, that have most of the properties (strength, speed, cunning, warm coats, and so on) that actually make a difference to the chances of survival. It is individuals that do the relevant surviving and reproducing. This suggests a very different, but prima facie more plausible, account of how genetic frequencies track the environment. They are, so to speak, dragged along by the selective processes that take place at the organismic level.

Apart from general reductionistic prejudices, there is a legitimate ground of resistance to such an organism-centered view of the process of selection. This is just that there *is* a real and crucial role for genetics in telling selective stories that might be obscured by the organism-centered view. The genetic properties of organisms do not *just* passively track organismic selection processes but rather strongly constrain such processes. Where, for example, organisms are undergoing strong selection for increased size, they are likely rapidly to run into genetically imposed limits on size. An organism-level model of selection requires not only the relative fitnesses of different phenotypes, but also some facts about the heritability of the trait in question. Given the complexity of the genetic influences on phenotypic traits, these facts may themselves be quite complex. It is often supposed that problems with the heritability of traits are just the exact obverse of problems with the fitness of genes. Genetic models solve the former at the price of having to deal with the latter, and vice versa for phenotypic models. Although this diagnosis has something to it—it locates real limits on the amenability of natural selection to systematic modeling, and locates part of the ground for that in the involvement of widely separated ontological levels—it exaggerates the similarity of the two problems. The heritability of a trait, though certainly only measurable as an average, is affected only by phenomena at one structural level, the genetic. For this reason, heritability is a more plausibly robust property of traits than fitness is of genes: different samples of a population can be expected to reveal approximately the same trait heritabilities, since these will not depend on environmental variables. Moreover, to the extent that the genetic basis of a property is stable, even though it may be very complex, its heritability will also be stable. Thus heritabilities offer some promise of being real (second-order) properties. The claim on which I would like to insist is the following: in most cases phenotypic selection models attribute the right properties to the right entities. Phenotypic traits are selected, and this is a wholly individual-level phenomenon. Traits are also more or less heritable; this is a property grounded wholly in their genetic bases, however complex. Genetic fitnesses, on the other hand, depending inextricably on both the genetic and the organismic, are misbegotten mongrel concepts, derived from reductionistic excesses.[22]

The general problem I see with population genetics can now be simply stated: Models of genetic changes in populations do not provide us with much illumination of the ways in which populations evolve because, at least to the extent that the change in question is due to selection, properties of genes are not, except in quite special cases, what cause changes in populations. Changes in gene frequencies, I suggest, are caused by changes at the phenotypic level that are often the result of differences in the fitnesses of phenotypes. The phenotypic topology over which selection can drive a population is determined by genetic possibility; but properties of genes do not push the populations across the landscape.[23] This picture suggests a rather different, modest but important role for population genetics. Much of population genetics does involve the exploration of possible (and impossible) genetic trajectories and equilibria for evolving populations. However, whether such results have much relevance to delineating the phenotypic possibilities open to real populations remains open to question. And even this more modest goal does nothing to circumvent the difficulties with genetic fitnesses, and thus it is not clear that population genetics can tell us anything concrete about the possible course of evolution by natural selection. To the extent that population genetics can help to serve this more modest goal of plotting the realm of the evolutionarily accessible, this possibility reinforces the main thesis I have defended about the real virtues of reductive understanding: that even when structural facts can tell us little about the actual course of events, they can be essential for understanding how a kind of phenomenon is possible.

The antireductionist morals to be derived from an examination of genetics apply widely. Just as classifications are driven by principles quite specific to the kind of phenomena they are intended to systematize, so explanations of the course of events in a domain of phenomena will require principles, generalizations, or models appropriately applicable to that domain. This is just as well, since such explanations will presumably be framed in terms of those generalizations. An understanding of the determinants of evolutionary change, I have argued, is a clear example of an explanatory project for which reductionism is not an appropriate strategy. But I have also tried to keep in focus the obvious fact that there are explanatory projects for which a broadly reductionistic strategy is appropriate. (Perhaps ontogeny is such a case.) The case of genetics, broadly construed, is admirably suited to explore the differences between such projects.

My skepticism about population genetics may seem, in a sense, outrageous. Surely the condemnation of a major scientific enterprise as, if not wholly misguided, of only tangential significance to the problems to which it is widely heralded as holding the key, should be grounded in a detailed examination of the empirical evidence rather than emerge from an a priori discussion. Surely, it may be said, a science should be given the benefit of the doubt in the absence of a very thorough investigation of its claims to empirical support. But this principle invokes too powerful a defense: any project that meets the minimal sociological criteria of scientificity is liable to be guaranteed immunity against any potential critic who does not both fully master the technical apparatus of the science and then review the empirical credentials of a substantial sample of its technical texts.[24] If serious doubts can legitimately be raised at a more general—even perhaps a priori—level, this demand constitutes a substantial penance for the critic already convinced that the project is either misguided or of little importance.

Substantial additional mitigation can be offered in the present case. That evaluating empirical adequacy is not a very central part of the business of population genetics is apparent from a cursory examination of population genetics materials. It might be said that this apparent neglect of the empirical is in part an artifact of my failure to address the data-collecting activities of population geneticists. And population geneticists really do sometimes collect data, even data on natural populations, as in the case of Lewontin and White's (1960) investigation of *Moraba scurra*. But such data serve more as grist for the theoretical mill than as evidence for the general legitimacy of the project of population genetics. Selection coefficients, after all, can always be *measured*. And claims suggested by the models of population genetics are not even intended to be literally true of real populations, but are idealizations applicable to the real world only when hedged by *ceteris paribus* clauses as powerful as they are implausible. Moreover, I am by no means claiming that nothing true can be said in the language of population genetics. My insistence that the real processes (in most cases) are at the individual level suggests that one should expect certain limited kinds of population genetics models to be empirically correct. Where a process of selectively driven change is facilitated by a sufficiently simple genetic basis to the relevant phenotypic properties, the genes will be (to take up an earlier metaphor) dragged along in such a way that the selective models will work equally well in terms of genotypes or phenotypes.

Sex, for example, is inherited in a relatively simple manner and in an overwhelmingly high proportion of instances can be unambiguously assigned to just one of two possible states. The latter fact makes phenomena at the individual level analytically tractable, while the former keeps the genes strictly in step with the selective processes. These peculiar circumstances make possible such celebrated analyses as Fisher's explanation of equal sex-ratios, and more subtle accounts of deviations from sex-ratio equality. Single locus, or at least genetically simple, determination of traits subject to a dominant selection process will provide cases, such as industrial melanism or sickle-cell anemia, that are satisfactorily represented at either the genetic or the phenotypic level. More generally, when transmission and selection coincide, models will reflect a kind of order that can be stated in terms of either genes or phenotypes.[25]

But as selection and inheritance diverge, and both selection pressures and patterns of inheritance become more complex, which way should we go? My suggestion is that we will do much better to focus on individuals, because that is where the causal processes are. Does this mean that I think we can get better models of evolutionary processes by thinking in terms of individuals? Not necessarily. I see no reason to think there is *any* mathematical structure that will usefully model many important aspects or episodes of evolutionary change. But at least the focus on individual organisms dispenses with the reductionist illusion that these processes can be understood as revealing the operation of hidden underlying mechanisms whose parts are the contents of the geneticist's beanbag. *If* there are systematic patterns amenable to mathematical modeling in the evolutionary process, most of them are likely to be found at the individual level, the level at which the vast majority of selective processes take place. But I am more inclined to accept a much more pessimistic conclusion. As nature diverges from the simple genetics and overwhelming selective forces of our simple paradigms, there is no reason to suppose that any mathematically intelligible model can realistically reflect isolable selective processes. The mechanistic philosophy implicit in population genetics provides a way of burying our heads in the sand on the edge of this epistemological abyss.

One further point should be made in defense of my criticisms of population genetics. If one does not think that evaluation of scientific enterprises should be left entirely to the scientific priesthood, it is presumably relevant to ask what benefits, whether practical or theoretical, accrue from a certain kind of scientific practice. I have been arguing that the theoretical benefits are few. I think that this claim is somewhat reinforced by the observation that there are few apparent practical benefits.[26] The most successful scientific enterprises can be seen as offering both general nontechnical theoretical insights and also identifiable practical benefits. To take two very obvious examples, Copernican cosmology offered a wholly new understanding of the major features of the universe; as subsequently developed it offered better predictions of celestial events, better calendars, and the like. Molecular genetics provides us with a thorough, and thoroughly materialistic, account of inheritance; it has also given rise to extraordinary technological achievements in the production of valuable biochemicals, and promises major medical achievements in the future. I suggest, by contrast, that the intelligent layperson who takes the trouble to learn about population genetics models will have contributed rather little to her or his general comprehension of the biological world;[27] and practical benefits neither have been delivered nor, even by the most ardent enthusiasts for population genetics, are they anticipated. Certainly no broad social benefits of the practice of population genetics argue against my theoretically grounded skepticism.

Genic Selectionism

A rather different attempt to connect the theory of evolution with genetics is the thesis of genic selectionism, which has been offered as a solution to the general issue of the "units of selection."[28] The units-of-selection problem is simply the question what entities (genes, individuals, groups, species, and so on) are in fact selected in the course of evolution by natural selection.[29] Much of the controversy derives from the thesis proposed by G. C. Williams (1966) and best known through the work of Dawkins (1976, 1982), that the unit of selection is always the gene. On the whole this thesis has not been warmly received by philosophers.[30] Recently, however, it has been defended vigorously by Kim Sterelny and Philip Kitcher (1988).

Sterelny and Kitcher defend their position by combining a pluralistic metaphysics and epistemology with monistic methodological goals. The pluralistic metaphysics, which they relate to instrumentalism (p. 359), is the denial that there is any determinate target, such as genotype or phenotype, at which natural selection aims. The pluralistic epistemology that they advocate is that biologists "take advantage of the full range of strategies for representing the workings of selection" (p. 360). Somewhat surprisingly, given all this tolerant pluralism, the chief virtue they claim for genic selectionism is its generality—that, unlike any higher-level units, genes are always available for modeling a selection process. This advantage is established by appeal to a range of cases discussed at length by Dawkins (1982). In one class of such cases the selective processes appear to take place strictly at the genetic level. The classic example is that of segregation distorters, genes that manage to get themselves into more than half of an organism's gametes by subverting the process of meiosis. Even though these are often harmful to the organism as a whole, they may nevertheless be selected. Dawkins puts even greater emphasis on what he calls "extended phenotype" cases, in which a property selected is, in some sense, external to the organism undergoing the selection process. These include such things as nests, webs, or dams, constructed by organisms for various purposes, and the behavior of organisms of other species consequent on manipulation by organisms of the species undergoing selection, as, for example, by cuckoos and cowbirds. However, in these cases Sterelny and Kitcher acknowledge that it is perfectly possible to model such processes in individual terms, specifically in terms of the behavior that produces the artifact or deceives the victim, so that what is at issue is the advantage of also being able to do so in genetic terms.

The theoretical motivation provided by Sterelny and Kitcher for espousing genic selectionism therefore seems rather slight: it provides the only unified perspective that can accommodate clearly gene-level selective phenomena. I take it, however, that the main point they wish to make is that the genic perspective is always available; and thereby they express their dissent from what they identify as the alternative position, that there is a hierarchy of possible targets of selection, and that adequate representation of a selective process requires identification of the correct one.[31] This constitutes a belief in a pluralism of process as opposed to the pluralism of representations advocated by Sterelny and Kitcher.

It will be clear that many of my sympathies lie with the pluralism of processes, since I have denied that selection need involve any intelligible patterns at the genetic level. On the other hand, this position gives rise to the problem of stating what determines that a causal process is taking place at one structural level rather than at another. Sterelny and Kitcher think that there are answers to such questions only when, as in the case of selection processes involving segregation distorters, there is only one level at which the process can be adequately represented. But although there are clear intuitions about how to provide definitive answers to some such questions, a systematic account of the issue is another matter. Such an account is offered by Elliott Sober (1984a, chap. 8). Unfortunately, I agree with Sterelny and Kitcher that the account of causation involved is untenable.

The general issues about causality that arise here will be discussed in detail in Chapter 9. The following two points, however, are germane here. First, although it may sometimes be possible to represent a selective process at one of several structural levels—indeed, I conceded above that this will be the case in those paradigm cases in which the genetic phenomena are tightly yoked to the phenotypic—why should this generally be possible? It appears that Kitcher and Sterelny do not think that it is possible to give an individual-level representation of the processes involving meiotic drive. Why should it not be equally impossible to give a genetic-level representation of the processes involved in, say, selection for a gaudy tail? Although I do not hope to prove that this is *not* possible, unless it involves simply the trivial mathematical claim that any sequence of gene frequencies can be fitted to *some* function, I know of no argument that it *is* possible. The intuition that it is, I suspect, is grounded in some kind of metaphysical privileging of the small over the large, a privileging that it has been the object of this and the preceding chapters to undermine.

The second point is somewhat more radical. An assumption that is perhaps shared by participants on both sides of the debate is that for any selective process there is at least some appropriate way of representing it. Of course, if "representation" is taken broadly enough to include simply describing the selective advantages of a trait, and claiming that that advantage contributes to the explanation of the trait's presence, I would agree that that is a possibility without which the process would not be a selective one at all. But it seems typically that something much stronger is supposed, that some model of some limited degree of mathematical complexity, and with some significant degree of applicability to cases of some general kind, is available. Although I can see the point of such an optimistic assumption as reasonable methodology, I can see little grounds for believing it in the present case. It might rather be prudent to bear in mind that evolution is a historical process of inconceivable complexity and particularity, and the same may be true of the selective subnarratives that I am currently considering. This is a theme that will also be taken up in more detail in Part III.

A central issue arising from this discussion remains unresolved. I have agreed that in those cases in which the genetic basis of a phenotypic property under selection is sufficiently simple, plurality of representation of the kind Sterelny and Kitcher advocate will be feasible. I have also suggested that in some cases, especially when the connections between the phenotypic and genetic levels become sufficiently diffuse, there may be nothing but a unique historical narrative. And finally, I find it plausible that between these extremes there may be cases in which a reasonably systematic representation is possible, but only at the appropriate level. It is generally agreed that meiotic drive provides cases for which this correct level is the genetic. And many, though evidently not all, theorists think that there are (probably much more numerous) cases for which the correct level is the phenotypic. But what exactly makes this true remains obscure. Although I am very sympathetic to Sober's proposal that this be understood in terms of causality, it is not obvious that our level of philosophical understanding of causality is such as to constitute this as much of a gain in clarity. At the risk of lending encouragement to the tyranny of the microstructural, we could perhaps use the less controversial genetic-level cases to see what would have to be true for an equivalent kind of case at the phenotypic level. At any rate, further progress with this topic must depend on a clearer philosophical view of causality, something to which I shall attempt to contribute in Part III.

7. Reductionism and the Mental

Whereas reductionism is generally perceived as at least a problematic doctrine when discussed by philosophers of science, philosophical discussions in other areas, most notably discussions related to the mind-body problem, often seem to take the truth of reductionism for granted. While a number of philosophers have stressed the difficulties, or even asserted the impossibility, of reducing the mental to the physical, they seem almost always to treat this problem as something peculiar, or anomalous, about the mental. Ghosts or suchlike are suspected of having found their way back into our machines. In this chapter I shall criticize two widely discussed attitudes to the mental that are broadly physicalist in intent. But the rejection of these positions here should not be taken to indicate something peculiar to the mental. On the contrary, it provides merely a further application of the general pluralism that has been advocated throughout this book.

The two physicalist theories of the mental considered here are of very different kinds. The first is straightforwardly reductionistic, whereas the second, based on claims about mental tokens rather than types, seems more naturally grounded in something like the weakly compositional materialism discussed in Chapter 4. I shall, argue, however, that neither is well motivated. My rejection of reductionist physicalism will, at this point, be no surprise. My reservations about the second kind of physicalism, on the other hand, will raise some rather different issues; in particular, it will raise some of the issues about causality that will be the topic of Part III.

Reductive Theories of the Mental

Several versions of physicalism propose general, systematic connections between the mental and the physical. Such theories are reductive not just in the sense of asserting the physicality of mental items but also in the sense of claiming some kind of nomological, or lawlike, connection between the two realms. (Theories that are reductive in the first, but not in the second, sense will be discussed in the next section.) The kind of reductionism that has been the topic of the last three chapters will require connections of some sort between mental and physical kinds. The recent history of physicalism begins with arguments for the application of classical, or derivational, reductionism to the mental, in classic articles by Place (1956) and Smart (1959). More recently such positions have been rather widely abandoned. Particularly influential in this retrenchment have been so-called functionalist critiques (Putnam, 1960, 1963; Fodor 1968). Very simply, what these critics claimed was that for something to be a mental state of a certain kind was for it to play a certain functional role in the cognitive and/or behavioral organization of its subject. But such a functional role could, for different kinds of mentally capable entities, be realized not only by arrangements of neurons, but equally well by silicon chips or even arrays of hydraulic tubes. Thus, mental kinds instantiate a different level of abstraction from physical descriptions of neural states; and therefore there could not possibly be a law of nature equating the one with the other.

Whether this observation really has much force against reductive physicalism is open to debate. Certainly, if these functionalist claims are true, statements of the form "pain is identical to brain state X," which figured paradigmatically in early discussions of reductive physicalism, must be false. However, it does seem to remain quite possible to identify pain for particular kinds of mentally capable creatures with whatever structural states happen to realize the relevant functional state for those particular creatures. And it might even turn out that pain-for-humans, say, turned out always to be realized by a particular neural state. David Lewis (1966), indeed, uses functionalist considerations to argue for an identity theory explicitly committed to classical reductionism.

Whatever the exact role played by functionalist analyses of the mental, it is clear that derivational reductionism has fallen out of fashion as a physicalist account of mental phenomena. It seems plausible that functionalists have, by directing attention to principles of classification for mental phenomena quite independent of the physical, made it increasingly incredible that mental and physical kinds should, even when restricted to a particular biological species, turn out to coincide. The possibility of different realizations of mental states for beings of different kinds applies equally, in principle, to different individuals of the same kind. Reductionists have recently tended to move to the opposite end of the derivational–replacement continuum, and espoused eliminativist positions according to which most of what we currently think of as mental will prove, in the end, to be fictitious. Such positions have been defended since early in the recent history of physicalism, by Paul Feyerabend (1963) and Richard Rorty (1965, 1970). Recently they have received much more extended attention, notably from P. S. Churchland (1986), P. M. Churchland (1981, 1985), and Stich (1983). In what follows, I shall focus especially on the defense presented by P. S. Churchland (1986).[1]

Churchland's account involves a conception of reduction as a relation between *theories*. Relations between theories have, of course, been a main focus of this and the preceding three chapters. In contrast, much of the tradition in the philosophy of mind has concerned the question whether mental properties (such as being in pain) could be identified with physical properties. In the context of derivational reductionism, it will be recalled, a condition of successful reduction is that the properties referred to in the reduced theory be identifiable, by means of bridge principles, with properties describable in the language of the reducing theory. Thus it is no surprise that identities between properties should have been the main focus of physicalist theories directed toward classical reductionist accounts of the mental. (Hence also the familiar term *Identity Theory* to refer to such physicalist accounts.) Since for an eliminativist the terms of the reduced theory will, in all likelihood, turn out to refer to nothing at all, this constraint is removed, and the eliminative reductionist need discuss only theories and general relations between them.

The next question is what theory is being reduced to what. Churchland's thesis is a speculation about the future of knowledge, not a report of the current state of things, so the theories involved need not actually exist. The putative reducing theory does nothing so vulgar, and is to be some future version of neurobiology. The nature of the theory to be reduced is rather more complex: it is "the integrated body of generalizations describing the high-level states and processes and their causal interconnections that underlie behavior" (p. 295). Such a "comprehensive Theory," which "delineates the psychological states and processes mediating perception, learning and memory, problem solving, cognitive mapping of an environment, and so forth" (p. 295), also fails as yet to exist. At any rate, these two future theories are the ones that should eventually be unified by a suitable reductive integration.

Churchland is very much aware of a problem with all this. Traditional worries about the mental have been concerned not with the relations between the as-yet-unknown constructs of some as-yet-unformulated theories, but with apparently quite familiar things, or states, such as pains, beliefs, memories, and afterimages. This lacuna leads Churchland to discuss a third theory, "folk psychology," the "theory" that articulates the relations between, and causal powers of, these more familiar items. And she diagnoses "virtually all arguments against reductionism" as depending on the "designation of some aspect of our common sense framework [that is, folk psychology] as correct and irreducible" (p. 299). Churchland adopts a radical response to such objections. If our commonsense conceptions of the mental cannot be adequately reconciled with mature neurobiology, it may well be "because folk psychology is radically misconceived. Ironically for its champions, folk psychology may be irreducible with respect to neuroscience—irreducible because dead wrong" (p. 384). An example of this strategy is Churchland's treatment of the familiar Cartesian argument from the indubitability of awareness. One reason for rejecting such an argument is that "there may be no such thing as awareness" (p. 309). Awareness is simply one concept in our folk psychological theory, and more-accurate theories may have nothing even equivalent to it.

In summary, the real scientific action is to be expected with the coevolution of a grand Theory of the internal causal determinants of behavior and a neurobiological theory of the brain. Meanwhile our current primitive folk conceptions of the mental will compete with these more modern rivals as best they can. If they prove to be too far from the truth they may go the way of phlogiston and the ether. Perhaps they will absorb and accommodate the new science sufficiently to retain some practical usefulness. At any rate, as long as we conceive of our understanding of our mental lives as constituting a primitive theory, we cannot be too concerned about what will happen to it as more-sophisticated theories are developed.

Any attempt to evaluate this story will require some attention to how the term *theory* is to be understood. Although there has been a great deal of discussion of what is involved in the replacement of one scientific theory by another, it is quite doubtful whether this discussion has provided any parallels for the hypothesized replacement of the science of psychology by the scientific study of the brain. It is even more doubtful whether it is legitimate to apply the term *theory* to so-called folk psychology[2] in any sense in which it could even make sense to imagine its replacement by a technical scientific theory. This prospect, finally, stretches credulity past its breaking point when it is realized that folk psychology is to be supplanted not by perhaps highly intentional psychological accounts of the cognitive or affective realms, but by an austerely physical account of the architecture and chemistry of the brain, since all more generously endowed theories are themselves ultimately to be replaced by some theory of this austere character.

Paradigms of scientific replacement in the history of science are episodes such as the decline of phlogiston theory in response to the rise of Lavoisier's primitive atomic chemistry, the replacement (at least in principle) of Newtonian mechanics by Einsteinian relativistic mechanics, and perhaps the rejection by educated opinion of theologically based accounts of biological origins with the ascendancy of broadly Darwinian accounts of evolution. Could any model drawn from such examples be applied to the anticipated colonization and eventual extermination of every other part of the science of psychology by the imperialist aggression of neurobiology?

An immediate problem is to decide what is to be included in the domain of psychology to which this replacement is to apply. Although we can exploit the advantages of nonexistence and assume whatever we like about future neurobiology, in the case of psychology we are more constrained by the need to envisage something that at least addresses the major issues of present concern to psychologists. Psychology is a highly diverse field, encompassing topics ranging from pattern recognition and the differentiation of kinds of memory, through reasoning strategies and spatial abilities, to the determinants of sexual preference and the effects of a disadvantaged background. Whereas some of these topics seem plausible targets for insights gained by structural investigation of the brain, others seem much less promising. Churchland recognizes this point and offers contrasting examples of cases that she thinks will, and will not, engender skepticism about reductionism. An example of the latter is:

1. If two stimuli, such as two bars of light or two circles of light, are presented in a properly timed succession, the scene is perceived as *one* object moving in space.   (p. 296)

More likely to encourage the antireductionist will be cases such as the following:

2. When someone has an opinion on an emotionally significant matter, conflicting evidence is treated as if it were supporting, impressions formed on the basis of early evidence survive exposure to inconsistent evidence presented later, and beliefs survive the total discrediting of their evidence base.   (p. 298)

Churchland attributes the difference between these kinds of case to the fact that generalizations of the latter kind concern representations. I must confess to discovering myself as the stereotypical antireductionist in that I agree which set of examples is likely to resist reductionism, although I am less sure how central to this difference is the presence of representations (perhaps because I am insufficiently certain what representations are). Central to the way

I do conceive the difference is something I have emphasized several times in the preceding chapters, the difference between an understanding of *how* an entity does what it does, and an understanding of *what* it does. A reductive approach will generally be relevant to questions of the former kind, but only tangentially so to questions of the latter kind. I do suppose that the generalizations that Churchland conceives as involving representations are often those that concern what people do. (Human) neurobiology, I believe, is a science concerned largely with how people work rather than with what they do;[3] psychology has some important concerns with both kinds of question; and folk psychology is mainly concerned with understanding what people actually do. Clearly, if this is correct it will show how confused is the expectation that folk psychology might be replaced by an activity that in fact has quite different goals. I shall elaborate on this claim momentarily. But first I want to say more about the hypothetical replacement of (real) psychology by neurobiology.

Despite all the complexity that has been discerned in post-Kuhnian accounts of scientific change, it remains possible, when we consider the paradigm replacements mentioned above, to discern both fundamental points of disagreement between the replaced and the replacing theory, and points of agreement that provide a background against which the significance of the disagreements can be assessed. Although the ramifications of such a replacement may unfold over a substantial period, it is also possible to see these fundamental disagreements as determining approximately the range of antecedent beliefs that are threatened by the process of replacement, namely, the beliefs that are premised on those points of fundamental disagreement. The success of Lavoisier's chemistry determined the correct approach to a fairly well-defined set of questions about the properties of material substances, and did so by rejecting an equally well-defined set of beliefs on such matters grounded in the rejected phlogiston theory. The rejection of Newtonian mechanics involved (or at least was for a time thought to involve) the repudiation of absolute space, and of any cosmological hypotheses that assumed the existence of absolute space. The triumph of Darwinism, finally, involves the rejection of divine creation in favor of a naturalistic account of biological origins. Insofar as we accept this replacement we can no longer, for example, account for the design of flowers in terms of their beauty or their aptness to convince us of the existence and benevolence of the deity. However little we still understand the origins of sex, it cannot be just that God thought it bad for the man to be alone and created an helpmeet for him. On the other hand, a great deal of functional analysis of the features of organisms is unaffected by whether the functionality is attributed to a process of natural selection or to a benevolent designer. In all these cases there is at least some continuity in the kinds of questions that are to be addressed by successive theories.

I have been deliberately somewhat vague in the preceding discussion of theory replacements. Although it does seem to me that the immediate consequences of abandoning the phlogiston theory are reasonably clearly delineated, more or less diffuse connections between different areas of knowledge make the ultimate consequences of such a replacement much less certain. And the appropriate scope of Darwinism is still debated, especially in relation to the social sciences. The reason I am not concerned about this vagueness is that I do not think that anything parallel, vague or otherwise, can be said about the alleged replacement of psychology with neurobiology.

The context in which eliminative materialism arises makes it clear enough that the supposed fundamental disagreement between the replaced and the replacing theory is the ontological one over the truth or falsity of physicalism. But the problem with this is that it is quite unclear that *any* parts of contemporary psychology are in fact committed to the falsity of physicalism, still less to the truth of dualism. I doubt whether any respectable part of psychology is committed to *any* ontological position at this level of abstraction. No doubt psychologists are committed to the existence of many more or less theoretical entities involved in the theories of their subdisciplines. They may be committed to more or less dubious internal representations, or to repressed desires and Oedipus complexes. But such commitments will presuppose some general ontological theory only on the basis of a wholly bogus trichotomy: either such entities are really physical (neurological); or they are made of some other, nonphysical stuff; or they do not exist. The practice of psychology is surely better understood on the following hypothesis, which it has been my aim in these chapters to make plausible: if there are representations, or Oedipus complexes, or whatever, then insofar as they are made of anything, they are made of physical stuff. But their role in our intellectual economy has little or nothing to do with their being made of anything, so that the claim that they are physical entities is wholly misleading. The fact that no serious part of psychology considers, or should consider, itself committed to nonphysical mentalism makes it quite unclear what areas of psychology are to be replaced by neurobiology. It is perhaps a credit to Churchland's intellectual integrity that she adopts the only plausible response to this problem: everything, including common sense. This strategy will, however, as I hope to show finally with the discussion of folk psychology, prove fatal to the project.

One further point needs to be made about the relations between psychology and neurobiology. The discussion so far has been conducted entirely in terms of the replacements of theories and beliefs. But this suggests an extremely impoverished view of a science. There is surely more to a science than a set of statements listing what is currently believed true by its practitioners.[4] A science includes, as well as some agreed-upon statements about the subject matter of their science, such things as sets of questions agreed to be worth attempting to answer; experimental techniques, and instruments operated in particular ways; and standards for evaluating the work of other scientists. Moreover, a science as an institution includes a hierarchical organization of practitioners and a system of credentials that determines position in the hierarchy, a set of sites where the work of science is carried on, and so on. I do not propose to look in detail at how these various aspects of a science might fare under a Churchlandian replacement. My point is rather that the kinds of replacements I have offered as paradigms are generally conceived as parts of the progress of a particular scientific project (loosely conceived). This conception requires that some significant proportion of the constituents of a science continue across the replacement. Clearly, this is the case for the examples I have offered. Some physicists were converted to relativistic mechanics, and biologists to Darwinism (see Hull,

1989, chap. 3). Such changes are usually consistent with continuing to address (roughly) the same questions with the same techniques and in the same institutional setting. It is hard to see how any of this background could survive the change that Churchland hypothesizes. It would be hard for a behavioral psychologist to apply for a comparable position in a neurophysiology department, and if somehow she were successful in such an application, her rats would presumably be removed from their now useless mazes and dissected. What all this suggests is that the appropriate parallel for Churchland's replacement is not with anything like the cases I have considered, but rather with the absolute abandonment of an entirely misguided practice. Our attitude to psychology when we look backward from Churchland's scientific utopia will not be, as is the case for Newtonian mechanics or even a good deal of functionalist biology carried on with basically creationist background assumptions, as a valuable though ultimately inadequate step on the scientific voyage, but rather as a complete mistake. Our parallels should be with astrology or phrenology. Whereas this strikes me as an unduly harsh judgment on psychology, it is absolutely untenable in regard to the so far much more successful human practice, whether or not it amounts to a theory, of nonscientific talk about the states of mind of ourselves and others.

The contrast between folk psychology and the imaginary future neurobiology provides the extremes of the contrast I have emphasized between how humans are able to interact with the world in anything like the complex ways in which they do, and what precise actions are carried out by particular individuals. Since the latter is unlikely to be much illuminated by neurobiology, neurobiology can hardly provide an adequate replacement for folk psychology. To illustrate this point, let me return to Churchland's example of a psychological finding unlikely to inspire reductionist fervor, the case of emotionally laden beliefs (generalization 2).

It seems to me an open question whether, in the abstract form given, this generalization is likely to be explained to any degree by neurobiology. If this was found to be an extremely robust cross-cultural universal we might reasonably suspect that it represented a characteristic bias in human reasoning powers, and this

is something that might be explained, at least in part, by elucidation of the neural substrate of such reasoning powers. But consider, instead, the question what are the matters that are likely to
be emotionally significant, a question that must be answered if we
are to have any idea to what situations the more abstract generalization should apply. Suppose, for instance, we had rather the
generalization:

> 3. Spouses who believe that their partners are unfaithful tend to
> ignore evidence to the contrary.

Since we are familiar with a cultural norm that marital infidelity
is an emotionally laden matter, it is easy to identify this as,
roughly, an instance of the broader generalization. But clearly the
latter, whether or not it might be a direct reflection of some
neurological mechanism, even if it may *contribute* to the explanation of the narrower generalization, is quite insufficient for any
such explanation. A sufficient explanation requires, for example,
some explanation of why such suspicions should be emotionally
charged; and that, in turn, requires some account of what it means,
in a particular cultural context, for two people to be related by
marriage, something that varies greatly from one culture to another. Finally, if we are considering the *idée fixe* of a close friend,
neither of these generalizations would be of much relevance, and
we would rather appeal to quite specific knowledge about that
person's temperament, relationship to her husband, and so on.

This example leads to the heart of the problem with the idea
of replacing folk psychology, a problem that parallels objections
that I have identified with other reductionist projects. The concepts of folk psychology relate to the subjects to which they apply
with regard not solely or even primarily to their internal states,
but to their interactions with their environment, especially their
social environment. As I stated it in the discussion of genetics,
these concepts look upward rather than downward (or perhaps
better here, outward rather than inward). From a certain perspective, this is perfectly obvious. The present example concerns beliefs (a subclass of Churchland's "representations"), and beliefs are
identified in language. One need not embrace a very subtle or
controversial version of the Private Language Argument to ac-

knowledge that language is in some sense a social phenomenon. Thus whereas abstract claims about beliefs, such as (2), might possibly reflect internal structure, when the beliefs are made more concrete in (3), contextual social factors that go beyond the internal structure of the individual are inevitably involved. This is all well-trodden if still controversial ground, and I do not propose to go over it here in any detail. The main point I want to make is just that for anyone who agrees with my intuition that a belief cannot be understood in a purely solipsistic way, the argument of the preceding three chapters is intended to show, *inter alia,* that this intuition is ontologically unproblematic.

The contrary intuition is that whatever mental properties pertain to an individual, they must do so on the basis of some structural features of that individual, even if this is so in no stronger sense than that of supervenience. But I hope it is clear by now that there is no need to insist on even such a weak physicalism. Beliefs, often at least, explain actions. Actions, again often, take place in social contexts that have much to do with determining what kinds of actions they are. (Consider, for example, the vast range of different things that might, in different circumstances, be accomplished by the simple action of uttering "I do"). The reductionist will insist that the social is merely a product of the diverse behaviors of the many individual agents by which it is constituted. But it is equally possible, and perhaps more natural, to think of the social and the individual as each constituting partially autonomous, causally efficacious, domains. Sometimes we can explain social phenomena in individual terms, and sometimes individual behavior should be explained by appeal to social factors. And as with my claims in the previous chapter about phenotypic causes of genetic change, when I speak here of explanation, I do mean *causal* explanation.

The gulf between folk psychology and futuristic neurology can be made even clearer with a rather different kind of example. P. M. Churchland (1985) offers wine tasting as an example of an activity in which the subtle discriminations of the expert reflect, and might in principle be replaced by direct reports of, the neurological consequences of the contact of the wine on one's palate. Instead of reporting on the hints of raspberry, might we then more accurately announce that neuron 6809D was firing? Picking up at random a bottle of wine, I read on the back that "though the structure is firm, the wine is well-rounded." It seems to me, to say the least, unlikely that there is some neuron or set of neurons that fires only when my palate is exposed to a wine of a well-rounded vintage. But even if this were, improbably, the case, the report of those firing neurons would not adequately replace the designation of the wine as well-rounded. The term *well-rounded* is not merely homonymous with a word that refers directly to a circular shape, and metaphorically to an education or a fine meal. No one, I imagine, supposes that there is some part of the brain stimulated equally by apples, fine vintages, or a well-rounded education. But it is the point of such descriptions of wine to suggest connections with a range of different experiences, not to serve as an instrument of chemical analysis, something that would be better achieved by pouring one's Château Margaux into a chromatograph. When we think of the description of wines as romantic, banal, presumptuous, adventurous, and so on, this point becomes even more obvious. It is in the public and social space of language, not the dark and private interior of the cranium, that the relevant set of associations is to be found. In the words of Robert Louis Stevenson, blazoned over the road that leads into California's Napa Valley, "And the wine is bottled poetry."

This example begins to expose the fatuity of the proposal to replace folk psychology with neurobiology. I suggested above that folk psychology was a much more successful human practice than any scientific psychology. What I had primarily in mind was that folk psychology provides us with the entire resources of language for making subtle distinctions between an indefinite array of possible mental states. This is one of the main things that language is for, and one of the things that language users, especially the most talented users, are most skilled at applying it to. One should be astounded and even appalled at the scientistic arrogance that supposes that the dry and highly specialized technical terminology of an esoteric subfield of science should supplant the instrument developed over centuries by the efforts of Shakespeare, Dante, and Dostoevsky—to speak only of the West—and many millions of others to describe the subtleties of the human mind.

At the root of this arrogant assumption there remain philosophical hypotheses that cannot yet be dismissed. The Complete Theory of the universe must be assailed from at least one more direction, specifically, the idea of a complete and seamless causal nexus. The idea that there is in principle a complete causal story to be told about every event, and consequently that whatever different causal principles there may be, they must all be fully and additively combinable insofar as they apply to the same events, I have referred to as the thesis of causal completeness. I argued in Chapter 4 that such a thesis provides the basis for the most compelling argument for reductionism. This thesis will be the main target of Part III. For the remainder of this chapter I shall consider a quite different kind of physicalist perspective on the mental. It will turn out that the idea of causal completeness is deeply implicated in this strand of physicalist thought.

Token Physicalism

Prior to the resurgence of reductive physicalism, physicalists persuaded of the obstacles in the way of the reduction of the mental tended to move toward so-called token-token identity theories. Such theories, generally explicitly opposed to the possibility of general correlations between mental and physical types, argued instead for the thesis that every mental item must be identical with some physical item. (This view may be referred to, more simply, as token physicalism.) Such a position may be seen as a retreat to a kind of compositional materialism: mental things must at least be *made of* physical entities (in the sense defined in Chapter 4). However, the thesis was presumably motivated by a version of compositional materialism that incorporated a strong assumption about the scope of materialism. For, prima facie, mental entities— if, indeed, the mental can even be said to constitute "entities"— would not seem to be made of anything. The claim that they are in fact made of physical entities was presumably, therefore, motivated by the idea that compositional materialism provided a criterion of the real. Here I shall not attempt to refute token physicalism, but only to criticize one very influential kind of argument that was offered in its support. The flaws in this argument are, I think, very revealing of the underlying intellectual motivations of physicalism.

The argument in question for the token identity theory is presented by Davidson in his paper "Mental Events" (1970). The argument rests on the following premises:

1. There are causal interactions between mental and physical events.

2. Whenever two events are causally related, there must be some description of those events under which it is a law that an event satisfying the first description will cause an event satisfying the second. (This is essentially an alternative statement of what I referred to above as the thesis of causal completeness. Clearly, such a formulation does not allow that laws might abstract from any causally relevant features of the situation described.)

3. There are no psychophysical laws.

The argument then runs as follows. Take any instance of causal interaction between a mental and a physical event. According to the second premise there is some law that, under some description, these events instantiate. Since the law in question cannot be psychophysical, it must be physical. (One might wonder whether it could be psychological. In fact Davidson does not appear to believe that there could be any psychological laws.[5] But even if one does not share Davidson's skepticism about the existence of psychological laws, one must admit that it will often be quite implausible that the physical event in question could be mental—for example, when the interaction is perceptual—so this alternative will frequently not be available.) Therefore, there is some physical description that the mental event satisfies; which is to say, it is a physical event. To generalize, at least all those mental events that enter into causal interaction with physical events must themselves be physical. (And if there are really *no* psychological laws, it also follows that all mental events that enter into causal relations with other mental events must also be physical.)

A noteworthy feature of this argument is that the strength of each premise is greatly augmented by its successor, in such a way that in context, at least, each premise becomes highly debatable.

I certainly do not wish to deny that there is interaction between the mental and the physical. On the other hand, it has frequently been a matter of debate whether such interaction should be described as causal. Presumably the issue here must turn on how wide or restrictive is the notion of causality in question; and the notion of causality introduced in the second premise seems a very restrictive one, so that the question whether psychophysical interaction should be considered as causal becomes a serious one. Moreover, the conception of natural laws that underlies the third premise, that these must at least be absolutely precise and universal, makes this notion of causality yet more restrictive.[6]

In fact, the notions of causality and natural law involved in Davidson's argument are so restrictive that it may legitimately be doubted whether anything whatever satisfies them. However, even if we do take these conceptions for granted, the argument is unsound. Let me try to illustrate this with a nonpsychological example.

Consider the singular causal statement "The explosion killed the rat" or, in the canonical form that explicitly relates events, "The explosion caused the death of the rat." What happened was that a closed container of gunpowder was detonated, causing a shock wave and a stream of fragments to move outward at high speed from the center of the explosion. The rat had the misfortune to intersect this stream and suffered all kinds of damage—more than sufficient, no doubt, to ensure its demise (or that of any other rat so treated). Suppose some law (in Davidson's sense) can be formulated guaranteeing the death of a rat so treated. (Strictly, we might want to say something like "the cessation of [some specified] biological function of any organism physically identical with the particular animal concerned," since *rat* and *death* are by no means precise terms.) But the antecedent of such a law would have to refer to a suitable rat-shaped cross-section of the aftermath of the explosion. And such an entity is neither an explosion nor identical with one. Consider another rat involved in these proceedings. This animal has the good fortune, despite being in the vicinity of the blast, of being protected by a large and well-anchored rock. Unfortunately, as well as being a physical event of the kind that killed the first rat, the explosion is also a chemical event, one aspect of which is the generation of a large amount of carbon monoxide. The second rat survives the blast only to be poisoned. So, just as surely, the explosion also killed the second rat.

Davidson's conception of a cause is of an event instantiating a precise universal law. Thus such an event will be both necessary and sufficient for the occurrence of an event satisfying the appropriate description of the effect. The difficulty presented by the example so far, then, is that what, in either case, was necessary and sufficient for the death of the rat suitably described (if, indeed, anything was) was an event having certain causally relevant features and lying some distance along one of the many causal chains emanating from the explosion. Even granting that the unquestioned causality present in this situation licenses the assumption of some underlying physical law, and that this in turn shows the existence of such an event lying on some such causal chain, it would plainly be mistaken to identify this event with the explosion. But this appears to be exactly analogous to the identification involved in Davidson's argument. Indeed there is no reason to suppose that an event physically indistinguishable from the actual explosion will be either necessary or sufficient for the occurrence of the event nomically linked to the death of the rat. That it is not necessary is clear from the case of the first rat; for at most only some part of the explosion is causally necessary for the production of the rat-shaped cross-section involved in the rat's destruction. In other respects the explosion might have been quite different. To illustrate the lack of sufficiency, consider the second rat. If the wind had been blowing furiously in the direction from the rat to the explosion, let us suppose, the rat would have survived.

A natural response to the last point would be to suggest that strictly it was not just the explosion but also the particular direction of the wind that caused the death of the rat. But if this reply is interpreted as meaning that it is not "strictly" true that the explosion caused the death of the rat, then there is no reason to suppose that everyday statements of psychophysical causal interaction are "strictly" true. If the "strict" truth of a singular causal statement turns on whether some event referred to by the given description of the cause is also describable in such a way that any event so describable is by itself sufficient for the relevant effect, then nothing short of the most detailed scientific investigation could serve to establish the truth of such a statement; and there is no reason to suppose that there *are* any true psychophysical interaction statements in *this* sense. (Of course, it has been well understood, at least since John Stuart Mill, that the truth of a singular causal statement requires nothing of the kind.[7] The additional requirement of the presence of oxygen, for instance, does not make it untrue that my cigarette end caused the burning of the barn.) The most, then, that we might have good reason to believe is that a mental event formed part of the causal nexus responsible for a given physical event (or vice versa). Clearly, no identification of anything with anything else can be inferred.

A similar argument has been propounded by Christopher Peacocke (1979, pp. 134–135), which runs as follows. Consider a mental event, say a pain, that causes a physical event, say the withdrawal of your hand from a hot kettle. (Presumably we should more strictly say "the occurrence of a pain" since by no standard is a pain an event.) Let this pain event be c_ψ and the effect be e. Suppose also that there is some event c_ϕ that occurs in your nervous system and is the cause of c_ψ. Suppose also that we know, by empirical investigation, that c_ϕ is the sole and sufficient cause of e; that is, we have a complete physical explanation of how c_ϕ causes e. Suppose finally that c_ψ is not identical with c_ϕ. The only possibilities are (1) that c_ψ and c_ϕ are jointly sufficient but also individually necessary for e, or else (2) that e is overdetermined by c_ψ and c_ϕ. But (1) is eliminated, since it contradicts the hypothesis that c_ϕ is by itself sufficient for e. (2) implies that the withdrawal would have occurred even if the pain event had not. And this "we normally take to be false, and it is not clear why we should change this belief" (p. 135). This constitutes a *reductio* of the hypothesis that c_ψ and c_ϕ are not identical.

In summary the argument seems to come to this. Assume that in such a case we could come to have a complete physical explanation of the effect. What then are we to say about the putative mental cause? It cannot, *ex hypothesi,* be an additional necessary condition for the occurrence of the effect; but since we do not want to deny that it is necessary, we must conclude that it is identical with the condition that we do know to be necessary.

My first objection to this argument is that I do not believe that it is true that the pain is necessary for the effect. Pains, I take it, do not occur unfelt; so patients under anesthesia do not have pains (presumably that is one of the points of anesthetics). But anesthetized patients can exhibit withdrawal reflexes. In fact, surgeons operating on anesthetized patients use drugs such as curare precisely to prevent such reflex contractions. Thus the argument against overdetermination is unsatisfactory. A possible reply would be to deny that pains cannot occur without being felt. However, the argument is still vulnerable to an objection analogous to that which I presented against Davidson's. A cause is not generally assumed to be the sum total of all the conditions necessary and sufficient for its effect. For the case of the pain I have already argued that it is not necessary for the given effect; it is even clearer that it is not sufficient. Perhaps the hand is glued to the kettle, or the victim is a believer in self-purification through suffering. But since the hypothesis was that the physical event c_ϕ *was* necessary and sufficient for the effect, it cannot be identical with the pain. Alternatively we could relax that assumption. But then the most that we could show was that $c_\psi + C_1 = c_\phi + C_2$, where C_1 and C_2 are further necessary conditions. But from that it would obviously be fallacious to infer that $c_\psi = c_\phi$. (Strangely, Peacocke admits that he is prepared to substitute the expression "is a part of the cause of" for "causes" [p. 134, n. 19] but does not notice that this would entirely undermine his argument.)

A likely response to this objection (as to the similar objection to Davidson's argument) is that even if it would be a mistake to identify the mental event with the physical event indicated by this hypothetical neurophysiology, the argument still establishes that the mental event must be some part of the physical event that is fully necessary and sufficient for the effect. Indeed, this possibility seems a striking omission from the alternatives considered in Peacocke's putative *reductio*. To deal adequately with this response would require a more detailed discussion of the nature of events than would be appropriate to the present context. Here I shall only make some observations that should suffice to show that such a response would at the very least face severe difficulties. First, I have not so far questioned the assumption that the set of conditions necessary and sufficient for some effect can reasonably be described as an event. But actually this assumption is dubious, to say the least. An explosion is certainly an event. But even if the blowing of the wind in a certain direction is an event, is the sum of the explosion and the blowing of the wind really an event? (Is the sum of the leaning tower of Pisa and my wristwatch an object?) And if it is, should we also include negative conditions, such as the fact that the rat was not wearing armor, or that it was not struck by lightning at the same instant as the explosion, in our construction of such a composite event? These problems are somewhat concealed in Davidson's account by the fact that Davidson wishes to individuate events in terms of their causal relations. But then what the preceding discussion should have indicated is that this conception of events sits very uneasily with Davidson's account of causality. For the class of events distinguished by applying Davidson's account of events in conjunction with his theory of causality might well be wholly disjoint from the class of occurrences we denominate events in prephilosophical discourse.

A second difficulty is the following: given that we know that some wholly physical set of conditions is a sufficient cause of some effect, and that we know that some mental event is among the causes of that effect, is it legitimate to infer that we can identify the mental event with any determinate part of those physical conditions? This question depends upon subtle and difficult issues about the referential relations of different parts of our conceptual scheme, issues that I cannot attempt to resolve here. These issues tend to be obscured by the spurious suggestion that we are identifying a mental event with a physical event that has been given antecedently by some physical inquiry.

Let me return to Peacocke. With reference to the supposition on which his argument is based, that "we have a complete and wholly physical account of both the causal antecedents of c_ϕ and also of the causal route from c_ϕ to e in neurophysiological terms; this account completely explains how the event c_ϕ causes e," Peacocke writes: "These are strong assumptions that we could know only by detailed empirical investigation; but it is hard to believe that a scientist *could* not come to know them" (1979, p. 134; emphasis in original). This blind faith in the ability of hypothetical scientists to provide complete physical explanations for any imaginable occurrence is reminiscent of Churchland's imperialist conception of neurobiology. Needless to say, I see no reason whatever to suppose that a scientist could come to know any such thing. As I tried to establish in Chapter 4, to assume that every event is amenable to complete physical explanation is wholly to beg the question in favor of physicalism. At any rate, this assumption can hardly be a legitimate basis for an argument purporting to help establish physicalism.

The preceding discussion prompts one more general remark on the nature of causality. A contrast that emerged from my main example above is that whereas Davidson invariably discusses causality in terms of a relation between two events, what we saw in the case of the explosion was rather the production of numerous causal chains. The metaphor of a causal chain clearly suggests a determinate sequence of objects or events, and this in turn suggests the following diagnosis of the problem in Davidson's account. Suppose A causes C via some intermediate event B. On Davidson's view, presumably there is some law under which A and B (suitably described) fall, and similarly for B and C. But since the description under which B is nomically connected to A need not be the same as that under which it is nomically connected to C, there is no reason to suppose that A and C instantiate a law under any description at all.[8] This shows why even if Davidson's analysis were correct as between successive links of causal chains, it need not apply between the end points, and this, in turn, shows why it cannot be expected to apply to everyday singular causal statements, in which there is no a priori limitation on the length of the causal chain involved.

A more critical look at this metaphor, however, reveals that this line of argument completely undermines Davidson's analysis. The expression *causal chain* suggests a determinate sequence of links. But in fact, causal chains are typically not sequences of some specific number of events, but continuous processes. Davidson's analysis, if it applied at all, could apply only to chains of just one link. But since there is really no determinate number of links in a causal chain, there is no class of causal chains of just one link. So Davidson's analysis of causality cannot apply at all.[9]

I shall conclude here with a more speculative remark on the relation between science and philosophy as it is illuminated by the physicalist arguments just considered. It is hardly surprising that science has a central role in epistemology and metaphysics. "Science," construed broadly and preanalytically, means little more than whatever are our currently most successful, or even just influential, ways of finding out about particular ranges of phenomena. Indeed, there are probably many philosophers who would like to see traditional epistemology entirely superseded by philosophy of science. Since, as a self-confessed empiricist, I take epistemology and metaphysics to have a good deal to do with one another, I do not want to deny the relevance of science to metaphysics either. However, the relationship of science and metaphysics in contemporary philosophy is often an unhealthy one. First, much contemporary metaphysics has an almost fetishistic reverence for science. But while this may even be harmless if a very broad conception of science, such as that indicated just above, is assumed, quite the opposite is typically the case. Thus, second, the actual conception of science often assumed is extremely rigorous and restrictive, and quite inappropriate to many domains of phenomena. In the previous discussion of recent ideas in the philosophy of mind, I have tried to illustrate aspects of this inappropriateness, and hence of the intolerable tension between these two aspects of the way science is incorporated into contemporary metaphysics. Such abuse of an excessively rigorous and restrictive conception of science is part of what I mean by the (intentionally abusive) term *scientism*. In the final chapter I shall take up this topic more fully, with special attention not so much to the metaphysical excesses illustrated here, but rather to the potential ethical and political dangers of scientism.

III. The Limits of Causality

8. Determinism

Degrees of Causal Order

Probably the single most important factor underlying ordered, mechanistic views of the world is the assumption of an omnipresent and wholly regular causal nexus. As I discussed in Chapters 4 and 7, such a conception of causality is closely linked with reductionism. It is also linked with essentialism. A central idea behind essentialism is that the essence of a thing will determine to which laws of nature that thing is subject. Even opponents of type essences such as Locke have imagined that individual essences might serve such a purpose, although the epistemological difficulties in supporting such a view are obvious. Parallel to and connected with my arguments against essentialism and reductionism, in this chapter and the next I shall argue that the prevalence of causal order is much less than is often supposed. To begin this project it will be helpful to distinguish a number of possible degrees of causal order.

The highest degree of causal order is surely provided by the traditional picture of determinism. I shall discuss determinism in more detail in the next section of this chapter, but for now it is sufficient to conceive of a deterministic world as one in which everything that happens is fully necessitated by antecedent circumstances. As necessitation is a transitive relation, such determinism seems to imply that everything that happens was necessitated by the manner in which the world began; or, since there may have been no such beginning, that everything was necessitated by the state of things at any earlier stage in the history of the world.

Sometimes determinism is used to qualify not the world but a theory, as in the common observation that Newtonian mechanics is a wholly deterministic theory. In this sense determinism might be thought of as a much more local phenomenon. However, genuinely metaphysical determinism seems almost inevitably an all-or-nothing affair. This is simply because if there are any events not fully necessitated by antecedent conditions, enough ingenuity can usually discover a way for such events to interfere with the production of the supposedly determined events. This is especially relevant to the curious and anthropocentric doctrine that determinism applies to everything except (some parts of) human behavior. Presumably the "free" acts of humans will constantly derail the otherwise determined behavior of inanimate things. At the end of the next chapter I shall show how a more plausible account of the relation between the behavior of humans and of other things emerges from a thoroughgoing indeterminism.

The first departure from this extreme form of causal regularity maintains a complete and exceptionless causal nexus, but one in which the causal relations are generally not deterministic, but rather probabilistic. Perhaps some subclass of events will have a probability of 1, but this will be a special case. On such a view most events will not be necessitated; and the connections between states of the universe will be more diffuse the more they are separated in time, allowing the interposition of more contingent events. Such a position will tend to imply a perspective on history that is evolutionary in the contemporary sense, in contrast with an earlier sense of evolution befitting a deterministic universe, one that refers to the unfolding of a preexisting and predetermined pattern. The world I am now contemplating is one in which the reign of law is as universal as under determinism; it is just that the laws decree only the range of possible events and the (precise) probabilities of their occurrence. This view might be called probabilistic uniformitarianism.

Probabilistic uniformitarianism turns out to face some serious difficulties, which will be discussed in detail in the next chapter. To anticipate briefly one of the main problems, the probability of an event of a certain kind might well be affected to some degree by a very large number of factors, perhaps so large that the particular constellation of factors in a particular instance might typically be unique. This possibility raises insuperable epistemological difficulties in determining the probability distribution for a particular case, but perhaps also metaphysical difficulties: if there is typically no way of determining probabilities affecting a specific case, why should we believe that there are any such indefinitely precise probabilities? A possibility that could not be empirically excluded is that there are only probabilities relative to some set of specified factors, but no general convergence toward a precise probability as more and more factors are considered. This possibility raises the further prospect that the degree of convergence on precise probabilities might vary considerably from one kind of case to another. And if so, the degree to which a domain of phenomena proved amenable to inductively grounded investigation would prove variable and unpredictable a priori. Picking up the geological allusion of my previous term, and perhaps reflecting the attitude of some philosophers to this proposal, we might call this less orderly causal structure probabilistic catastrophism. The admission of probabilistic catastrophism opens up a continuum of possibilities as the degree and prevalence of convergence on determinate probabilities alter. The position of the real world on such a continuum would be strictly an empirical matter.

Finally there is the extreme possibility of complete randomness.[1] By this I mean that the probability of an event is quite independent of any antecedent events; it is not causally connected to anything, but just happens. Nobody seriously supposes that the universe is *entirely* random in this sense. But in a world characterized by probabilistic catastrophism randomness might occur in places. One interesting possibility, going one step beyond the case considered in the preceding paragraph, is that there might be kinds of events for which the consideration of more antecedent factors affected the probability of the event but in no consistent way. In this case the apparent causal structure would depend entirely on what set of factors one happened to consider, and would, therefore, be ultimately quite illusory. In such a case, if there were any productive strategy of empirical investigation, it would not be causal or inductive.

Having provided an outline of this hierarchy of successively less ordered kinds of causal structure, I can lay out my plan for this and the next chapter. In the present chapter I shall explain why a considerable number of philosophers have abandoned determinism in favor of probabilistic uniformitarianism, and endorse the reasons for this move. In the next chapter I shall argue that this is a highly unstable position, and that it inevitably collapses into probabilistic catastrophism. This, in turn, leads to the conclusion that the degree of causal structure in different parts of the world may be quite variable, and is in any case a strictly empirical matter. And this, finally, will provide an important perspective for the final two chapters, on the unity of science and the role of values in science. At the end of the next chapter I shall offer, for good measure, a solution to the problem of the freedom of the will.

The Meaning of Determinism

It is a commonplace that philosophical discussion of causality has, until quite recently, been conducted entirely in the context of the assumption of determinism. Even Hume, both the dominant figure in the history of such discussion and a notorious skeptic about the basis of our causal beliefs, appears to take it as obvious that for something to be a cause of a certain effect, every instance of it must be followed by that same effect: "we may define a cause to be *an object, followed by another, and where all the objects, similar to the first, are followed by objects similar to the second*" (Hume, 1748, p. 51; emphasis in original).

This definition hardly gives an adequate account of what is meant by determinism. Just how difficult it is to provide such an account can be gleaned from John Earman's brilliant but often difficult and highly technical treatment of the status of determinism in classical and contemporary physics (1986). Determinism is often thought of in terms of Laplace's demon, a being who is imagined to be able to measure the states of all the fundamental particles in the universe and to infer from those their states at any future time. The problem with Laplace's demon is that he is liable to convey the suggestion that determinism is an epistemological doctrine, as is revealed by the frequent association of determinism with the possibility of prediction. One philosopher who makes such an erroneous assimilation is Karl Popper (1982), whose position is succinctly discussed and criticized by Earman (1986, pp. 8–10). Predictability presupposes, in addition to the determination of future states by present states, first, that the present state be ascertainable with sufficient precision to permit accurate prediction, and second, that the mode of dependence be mathematically tractable. With regard to the second point, the apparent insolubility of the three-body problem drives an obvious wedge between determination and prediction, although perhaps we should assume that Laplace's demon has a solution to this and indeed to the N-body problem.[2] With regard to the first point, the assimilation of determinism and predictability has become particularly unfortunate with the rise of chaos theory, a branch of mathematics that deals with perfectly deterministic functions, in the sense that the values of the variables at time t fully determine their values at any later time t', but functions that are nevertheless indefinitely sensitive to the exact values of certain parameters. Thus if a physical system were correctly characterized by such a function, no measurements to a finite degree of precision of its parameters at a time would suffice to predict its state at a future time. Many physical systems, for example those in meteorology, are hypothesized to obey such functions, and thus to be both deterministic and in principle unpredictable.

What can safely be said about Laplace's demon is that determinism is a necessary but insufficient condition for the successful execution of its responsibilities. Any reliable prediction requires that some aspect of the present state of things be a sufficient condition for some future state. The kind of metaphysical order involved in determinism involves the invariable truth of this precondition of prediction. The main tradition of understanding determinism has involved a concern with this metaphysical condition. However, stressing that determinism is fundamentally a metaphysical doctrine, and hence independent of whether or not we can make particular kinds of predictions successfully, should not obscure the fact that *evidence* for determinism will tend to come precisely from our ability to predict the course of events. Kant, perhaps, thought that determinism could be deduced a priori from the possibility of knowledge. This strikes me as a wholly incredible project: it seems too easy to believe that we are actually in a world in which many phenomena are not determined by the antecedent state of things. To show that this possibility does not obtain requires empirical evidence; and the form of this evidence would be cases (of which *some*, no doubt, can be found) in which the present state of things enabled us to predict their future state. The perception of the thoroughly empirical status of causal relations is the insight that gives Hume his deserved place at the center of philosophical discussion of causality. It is curious, and perhaps unfortunate, that he nevertheless seems to have treated determinism as something that it was at any rate beyond human powers to doubt. History has proved him to be empirically mistaken.

Unashamedly metaphysical definitions of determinism can be found in the writings of many of the leading philosophers of the twentieth century.[3] Earman (1986, pp. 10–12) considers one natural approach deployed by Bertrand Russell (1953). Russell proposed that if a system is deterministic through some time interval t_1 to t_n, then there is some set of values of parameters e_i describing the system at t_i, such that the state of the system E_t for any time t in the relevant interval is some function of the values of $e_1 - e_n$ and $t_1 - t_n$. Unfortunately, as Russell himself noted, this approach leads quickly to triviality, since it is mathematically necessary that some function, admittedly perhaps very complex, exist to satisfy this condition. Earman remarks: "Combining Russell and Popper, we have the first intimation of the Scylla and Charybdis between which determinism is forced to sail: tack one way in defining determinism and determinism wrecks on obvious falsity; tack the other way and it wrecks on triviality" (1986, p. 11).

What seems to be needed to rescue Russell's approach from this triviality is some appeal to laws of nature, or perhaps to the uniformity of nature, which seriously—indeed uniquely—restricts the possible states of a system in light of its earlier states. The approach that Earman favors involves appeal to possible worlds: if two possible worlds agree in all respects at some time, then they agree at every time (or, perhaps, just at every future time). A similar definition is offered by David Lewis: "The prevailing laws are such that there do not exist any two possible worlds which are exactly alike up to some time, which differ thereafter, and in which those laws are never violated."[4] Skeptics about possible worlds will obviously have reservations about such accounts. However, they capture clearly the metaphysical intent of determinism, and they will be adequate for my main purpose here, which is to argue that there is nothing like determinism that we have any good reasons for believing.[5]

The Retreat from Determinism

Despite my assertion that deterministic assumptions have dominated philosophical discussion of causality, the possibility of an indeterministic understanding of causation has received some parallel attention. C. S. Peirce, notably, argued forcefully for the idea that laws of nature should be seen as statistical regularities (1972, pp. 174–190). The subsequent development of quantum mechanics has made such views more attractive, although it is not clear to what extent the currently dominant view that the behavior of elementary particles is irreducibly indeterministic[6] bears on the tenability of determinism at the macroscopic level.[7] Most twentieth-century writers on scientific laws and causality have given at least some attention to the possibility of probabilistic laws.[8] More recently, dating particularly from the work of Patrick Suppes (1970), some philosophers have defended a more generally probabilistic view of causality.

I must now attend to some fundamental distinctions, somewhat glossed over in the preceding discussion, between probabilistic (or statistical) laws, probabilistic (or statistical) explanations, and probabilistic causes. To begin with the first of these, whereas deterministic laws of nature are generally assumed to take the form of universal generalizations concerning the concurrence or sequence of events or of the instantiation of properties by objects, probabilistic laws replace the universal "all" in such generalizations with "$x\%$ of" for some $x < 100$. Thus instead of "All crows are black," which has sometimes been proposed as a paradigm of a universal law, something more plausible, such as "99.9% of crows are black," might be an example of a probabilistic law.

As so far described, probabilistic laws in no way entail that there are any genuinely indeterministic processes in nature. One might well suppose, for example, that *all* crows (or perhaps birds, corvids, or some lower taxon) with genetic feature B would have black feathers; and that 99.9% of crows had this genetic feature; and consequently that 99.9% of crows are black. The second premise is merely what Mill called a "collocation" (1875, bk. III, chap. 12), on a par with "75% of the coins in my pocket are dimes," which nobody would suppose was any kind of natural law. More familiarly, something of the sort is often assumed about cases such as "⅙ of the tosses of a regular-shaped die of homogeneous density are sixes." The idea here is that every toss instantiates, with sufficiently precisely specified parameters describing the initial conditions of the toss, universal laws of mechanics. That ⅙ of the sets of initial conditions would lead to a toss of six would then be explained, perhaps, by appeal to considerations of symmetry.

However we understand the underlying ontological status of probabilistic laws, there are natural ways of applying laws of this kind to the provision of probabilistic explanations. The general account of scientific explanation that remains, despite well-known difficulties, dominant in the philosophy of science, so-called deductive-nomological (or D-N) explanation, involves the idea that to explain something is to deduce it from statements of which at least one is a law of nature. If what is to be explained is a particular fact, further known particular facts will also be among the premises of the deduction. Various attempts have been made to extend the basic idea of D-N explanation to indeterministic phenomena. The best-known such extension was worked out by Hempel under the rubric "Inductive-Statistical explanation," generally now referred to as I-S explanation.[9]

In view of my fundamental concern with causality here, it will be useful to consider an explicitly causal paradigm of I-S explanation, which might be, at its simplest:

$$p(E_{t+1}/C_t)^{10} \text{ is close to } 1$$
$$C_t.$$

Therefore, $p(E_{t+1})$ is close to 1, or (E_{t+1}) is nearly certain. Superficially, this looks very similar to the D-N explanation:

$$C_t \to E_{t+1}$$
$$C_t$$
$$\therefore E_{t+1},$$

where the first premise is supposed to represent a universal causal law. These examples intentionally embody a standard conception of the relation between laws and causes, a conception that reaches a decisive statement in the work of Hume: C is a cause of E just in case there is a law of nature that whenever C occurs it is immediately followed by the occurrence of E. (Hence the arrow may be read either as a causal relation or as a material conditional.) Citing both such a law and the occurrence of the cause is then seen as providing the ideal causal explanation of the occurrence of the effect. The modification of these ideas seen in the model of I-S explanation cited is natural enough. The relevant law can be obtained by replacing "whenever" in the above definition by "almost always, when." Given that E does in fact occur, it is explained in just the same way in either case.

Despite the superficial similarity of such D-N and I-S explanations, it turns out that there are profound differences between the two.[11] One central difference is the following. If $C \to E$ (ignoring, for now, the time indices), then $C + X \to E$, whatever X may be. But if the probability of E given C is very high, it does not follow that the probability of E given C and X is high. To adapt an example from Nancy Cartwright (1983, p. 31), ingesting sulfuric acid might almost always cause death; but ingesting sulfuric acid while simultaneously ingesting caustic soda might seldom cause death. This disanalogy, however, is easy to exaggerate; for in stipulating that something is a deterministic cause we are implicitly assuming that all possible interacting or counteracting factors have been included in the description of the cause itself. So if we had taken ingestion of (a sufficient quantity of sufficiently concentrated) sulfuric acid as an example of a deterministic cause of death, we would have had to assume that any possible countervailing force, such as the ingestion of caustic soda, is specifically excluded. Thus X in the deterministic case cannot be anything at all. It cannot contradict anything implicitly included in C. These issues about interacting causes will be taken up in some detail below. For now, a quite different difficulty is more relevant. This is that once we take probabilistic laws seriously, the requirement that causality should show up in something close to perfect correlations between cause and effect seems quite unmotivated—or anyhow motivated only by the assumed parallel between D-N and I-S explanations.[12] Suppose that we have two fair dice, each of which is marked on just one side with a six. Suppose further that we are interested in just one kind of event, that when the dice land showing double six. It would appear that when an outcome other than double six occurs, this is fully explained by the fact that thirty-five throws out of thirty-six will produce this result. But clearly there is no possible I-S explanation of the case when the dice do land double six, since this event has only a probability of 1 in 36. But this asymmetry seems quite unintelligible. A full account of the behavior of the dice will predict just as surely that approximately one toss in thirty-six will land double six as that thirty-five out of thirty-six will fail to do so.

More obviously causal is a much-discussed example from Michael Scriven (1959, p. 480). Paresis is a form of tertiary syphilis and is suffered only by those who have passed through the earlier stages of the disease. However, most people who contract syphilis, even if they never receive treatment, do not acquire paresis. It seems obvious that prior infection with syphilis could be cited as a causal explanation of someone's acquiring paresis. Yet such prior infection does not make it very likely that a person will acquire paresis, so that this case cannot be accommodated by the I-S model.

The solution to these difficulties requires a much more radical break with the conception of explanation associated with the D-N model. The starting point for this reconception is the rejection of the assumption that explanation involves any kind of inference.[13] This alternative view replaces the notion of statistical inference with that of statistical *relevance* and immediately resolves the problem posed by the syphilis/paresis case. The fact that a person has suffered untreated primary syphilis is certainly statistically relevant to the likelihood of his or her going on to acquire paresis. To develop this idea further it will be necessary to consider more explicitly the notion of a cause. The basic idea that forms the starting point for such an analysis is that the presence, or the introduction, of a genuine cause must make the occurrence of its effect more likely than if it had not been present. Thus the occurrence of paresis is more likely, even though not very likely, in a person who has untreated syphilis. And the landing of the dice at double six is more likely when they are tossed than when they are not.

In light of this analysis, the basic form of a probabilistic causal law will be the following:

$$p(E/C) > p(E/-C).$$

A standard example replaces C with "smoking" and E with "contracts lung cancer" and proposes that the probability of acquiring lung cancer is, as a matter of causal law, higher for those who smoke than for those who do not. Or more simply, smoking causes lung cancer. Something like this appears to be quite unproblematically accepted in ordinary discourse as the interpretation of claims such as that smoking causes lung cancer. Only the very unsophisticated take the anecdotes about Uncle Albert, who smoked all his life, never missed a day's work, and was hit by a truck at the age of eighty-seven, as refuting this familiar causal claim. And it takes only slightly more sophistication to realize that lung cancer does not even need to be very likely for the individual smoker.

With some serious complications that will be discussed in the next chapter, I shall assume that this is the right way to think both about probabilistic laws and about the correlative *general* causal facts. It is less clear how useful the statistical relevance concept is in understanding particular instances of causal connection, and hence in explanation of singular events. (As is common practice I shall refer to such putative explanations as S-R explanations.) Suppose Harold, a lifelong smoker, dies of lung cancer. Did his smoking cause, and does it explain, his lung cancer? There are two problems here. First, it is obvious that unlike a D-N explanation, an S-R explanation does not show that the effect had to happen. In the same way, it does not show why Harold, but not Albert, acquired lung cancer. The correct response to this problem, I think, is to acknowledge both that Harold's lung cancer indeed did *not* have to happen and that there may be no reason why Harold became ill and Albert did not. Albert was lucky and Harold wasn't. This is surely just the kind of thing we expect if we take seriously the idea of truly probabilistic laws. From this point through the next chapter I shall be defending the idea of taking chance, or luck, seriously.

The second problem is how we know that such explanations are correct or, for that matter, what it is for such an explanation to be correct. A natural answer to this question in the deterministic, D-N, case is that if the explanation is correct, then the event being explained would not have occurred but for the facts cited in the explanation. This goes beyond the previously mentioned conditions for D-N explanation, in that it rules out cases in which the explanation is redundant, either because what is explained would have happened come what may (for example, explaining a man's failure to become pregnant in terms of his taking birth control pills),[14] or because the effect is overdetermined (lightning struck just as the match was thrown into the barn).[15] But with these somewhat technical qualifications, D-N explanations show both that the effect had to happen given the occurrence of the cause, and that the effect would not have happened if the cause had not. This is an attractive pair of criteria for a fully adequate explanation. Nevertheless, if it seems likely that there are laws of nature that are irreducibly probabilistic, these are goals of explanation that sometimes will be unattainable. I have already discussed the impossibility of achieving the first criterion. And the second is equally clearly inappropriate to a probabilistic case. Consider again the paradigm, smoking causes lung cancer. On the account of such laws just introduced, this means that the probability of lung cancer is higher if one smokes than if one doesn't. But clearly this is compatible with what is in fact the case in this example, that some nonsmokers are unlucky enough to contract lung cancer. So can we ever *know* that Harold's smoking caused his lung cancer, in the sense that had he not smoked he would have remained healthy? The answer surely must be no, showing that the second criterion must be abandoned. Unlucky though Harold may have been, he might have been unluckier yet: he might have acquired lung cancer even though he, unlike Albert, had never smoked.

The last examples may seem implausible in that they may seem to trade on my refusal to consider in detail how further background or contributing factors must be included in an adequate causal explanation. Perhaps there is always available a sufficiently detailed specification of the circumstances such that in those circumstances the cause is necessary for the effect. Postponing the discussion of background or interacting conditions to the next chapter, for now I shall simply assume that we have established the most specific relevant description of the cause in question.[16] There is surely a possibility that we will then have specified a set of conditions under which the putative cause is necessary for the effect. But there can be no reason to assume that this possibility generally, or even frequently, obtains. For why should the combination of these background conditions with the absence of the putative cause have a mysterious tendency to provide us with a deterministic *preventative* of the effect, which is what would be needed if the background probability of the effect is to be 0, and the cause thus necessary for its effect? If we are genuinely committed to the existence of probabilistic laws we are unlikely to anticipate the existence of such deterministic preventatives. The attractiveness of the intuition that I have just been criticizing seems to derive entirely from residual deterministic prejudices: when we have specified *all* the relevant circumstances we will find that the effect is either necessary or impossible. Often such prejudices will be fleshed out with the idea that all the relevant circumstances would provide us with some kind of mechanistic story. Certainly it is plausible that there should be some such story to tell in the example of smoking and lung cancer; no doubt there are intermediate steps, such as interactions between toxins in cigarette smoke and constituents of the cells in the human lung. But the existence of such finer-grained accounts is irrelevant to the fundamental question, since the issue is simply converted to that of whether these fine-grained causes are themselves deterministic or not.

The conclusion, if we take seriously the idea that the causal structure of the world might be irreducibly probabilistic, is that our explanations must often fall short of the desiderata just discussed, and, perhaps more disturbingly, there may often be no way of telling whether something proposed as a cause of some effect was in fact required for the occurrence of that effect. Explanation, and the identification of causes, may often be limited to distinguishing factors that influence the probability of the effect to be explained. If, as in the case of smoking and lung cancer, the probability is very greatly affected by the suggested cause, we have a persuasive explanation. But that is the most to which we can legitimately aspire. The attempt to reinstate the criteria of explanation just considered reveals only incomplete emancipation from the assumptions of determinism. It is time, therefore, to consider more carefully whether we are justified in abandoning determinism altogether.

The Falsity of Determinism

There are two very powerful reasons for rejecting the doctrine of determinism. The first is that it seems almost entirely, or perhaps entirely, devoid of empirical support. The second is that our most successful scientific theories describe a probabilistic rather than a deterministic world. In the light of these two facts it is remarkable that determinism should remain a live option. Apparently these claims are not generally recognized as facts, so in what follows I shall briefly explain why I take them to be such. Given the enormous history of discussion of this topic, my remarks here must remain less than comprehensive. In this section I shall also consider the two main strategies for connecting a philosophical account of determinism with an account of the causes or laws supposed to do the determining, and argue that neither is at all plausible.

Let me approach the problems with determinism by briefly considering a much more defensible interpretation of determinism, but one that is ultimately irrelevant to my present concerns. This is the idea that determinism is really not a metaphysical doctrine at all, but a methodological precept. Deborah Rosen (1982–83) begins a critical discussion of deterministic causality with the following example: "One would be surprised, upon testing one's yogurt, to find it soupy if, on former occasions, similar ingredients and procedures had made it thick and firm. We would assume, of course, either that we had done something different on the occasion with the unusual outcome or that there was a hidden variable at work which would account for the difference in issue" (p. 101). Rosen's claim, with which I fully concur, is that these assumptions might perfectly well be mistaken. Why then does the "of course" in the second sentence seem so natural and unexceptionable?

Part of the answer undoubtedly is that we have deep deterministic impulses or prejudices. But a more legitimate reason is that we are surely right in thinking that if, say, we were professional yogurt makers, it would be crazy to assume at once that our failed batch was just bad luck. We would *(of course)* look for some explanation of the difference in outcome in terms of a difference in process or ingredients. Investigation seems to require the commitment to look for an explanation when apparently similar situations produce different results. Moreover, in the case as described it is very easy to think of directions that such investigation should take: the anomalous yogurt might differ in quality of ingredients, intrusion of unwanted microorganisms, ambient temperature, humidity, and so on. If we assume the process has been very carefully controlled—with sterile containers, precisely regulated ambient conditions, and so on—we are more likely to consider seriously the possibility that there is no explanation beyond chance, although even then it would be rash to assume without extensive investigation that we had eliminated every possible cause of the difference. No antideterminist supposes that we should attribute all apparent differences in behavior to chance, or denies that it is generally appropriate to look for explanations of different outcomes. The crucial point is that we have no right to assume that such explanations are always to be found, or that the end point of our investigation will be the discovery of exceptionless generalizations.

A clear sense in which there *could* be no *decisive* evidence for determinism is that determinism claims that there are exceptionless universal laws of nature. Notoriously, the truth of such laws cannot be empirically established. However it is not part of my goal here to address the problem of induction. The relevant question is whether our experience of the natural world, extrapolated without worries about induction, should encourage us to believe in determinism. It seems to me overwhelmingly apparent that this is not the case. Some realms of phenomena that come to mind immediately in this connection are the subject matters of population and behavioral ecology, economics, meteorology, evolutionary biology, geology, and almost all of the diverse aspects of human psychology. I do not deny the *possibility* that deterministic systems of laws might someday be discovered in such domains; I claim only that nothing like such a system has yet been discovered in any of them, so that no evidence is yet extant that such should be expected. Progress in such fields, where any has occurred, consists rather in better conceptualizing phenomena and in discovering further (indeterministic) causal capacities.[17]

One objection to the preceding claim might be proposed by devotees of chaos theory. It has been suggested that meteorological phenomena, for example, are quite compatible with the hypothesis that the weather obeys deterministic laws, but laws that also exhibit the kind of instability characteristic of chaotic systems. If this were the case it might, indeed, be impossible to discover a set of deterministic laws governing such phenomena, because of the impossibility of making sufficiently (that is, infinitely) precise measurements of the state of the system. But this situation would be compatible with the system's nonetheless being a genuinely deterministic one. This objection is easily dealt with. My present concern is whether there is evidence that favors a deterministic picture of the universe. The objection offers an argument that such evidence might be unobtainable even if the universe were deterministic. Perhaps the objection succeeds in limiting the damage to determinism that follows from denying that there is any evidence for it, but certainly it cannot help to show that there really *is* evidence.

A more effective reply is to insist on the appeal to a different set of examples. If a sufficiently compelling set of examples that do favor determinism can be found, then perhaps some comfort could be drawn from an inductive argument suggesting that eventually more-recalcitrant cases will also be led into the deterministic fold. The most obvious example would be Newtonian mechanics,[18] the area of science that, notoriously, has done most to inspire modern determinism. But first, the insolubility of the N-body problem is sufficient to show that we lack evidence for the general applicability of deterministic laws to systems of bodies in motion. Second, as Nancy Cartwright (1983) has noted, laws such as those of Newtonian mechanics are not even true except when, as will almost never be the case, other forces, for example electromagnetic, are absent. Finally, we should generalize this last point to a much wider range of forces than those that predominate in physics texts. One of the reasons falling objects near the surface of the Earth do not obey the simple Newtonian account of the behavior of falling objects is that they are liable to be affected by winds.[19] If the weather is an irreducibly indeterministic system, then the behavior of falling objects near the surface of the Earth is itself indeterministic. Hence the division of our experience into apparently deterministic and apparently indeterministic systems is a two-edged sword. While there may be an analogical argument from the former to general determinism, there is a causal argument from the latter to general indeterminism. Global determinism can obtain only if it is wholly isolated from causal interaction with objects whose behavior is indeterministic. This simple point places a heavy burden of proof on the proponent of global determinism.

Perhaps more compelling than belief in deterministic systems of laws is experience of the extremely predictable behavior of certain objects, especially artifacts. It may seem hard to imagine how such objects are possible at all if they are not exploiting deterministic laws. We may begin by noting that in a certain sense this is truly the heartland of the metaphysics I am opposing, the metaphysics of *mechanism*. Since the strategy of mechanism is to take machines as the paradigm for understanding the world, it is perhaps sufficient for the refutation of mechanism to note that machines are, in the respects I am currently concerned with, not at all typical of most of the world.

There are a number of interesting questions to be raised about the nature of machines, but for present purposes two simple points should suffice. First, machines are never completely deterministic. However impressed we may be by the Toyota that runs completely free of problems for 100,000 miles, we know that it will not run for 1 million miles. And we know that some cars (even, I imagine, some Toyotas) do not come close to this ideal of predictable behavior. It is also true that methodological determinism is liable to seem at its most compelling here: we are completely confident that if a car doesn't run there is some reason for this. But a crucial part of why we believe this is just that when cars don't work we simply replace substantial interchangeable parts of them until they do. Cars can, in this sense, always be fixed. And we can, in the same sense, always give a reason why they failed—the distributor went bad, the piston seized, the engine exploded, and so on—identifying the part or parts that we have had to replace. Empirically, some—usually most—parts of any particular kind work; and definitionally, if all the parts are working, the whole machine will work. (Recall here the connection between determinism and reductionism.) Often exactly the same things can be said of the parts of the parts. But it does not follow that whether a particular part, or part of a part, is working is a deterministic matter. At some point we come to a level that may perfectly well be indeterministic. And indeed if this level involves phenomena such as metal fatigue, abrasion, corrosion, or other such gradual degradations of the properties of structural materials, the relevant phenomena are most plausibly indeterministic. Finally, in qualification of my claim that any car can be fixed, I might note that a sufficiently egregious lemon might, by the time it was fixed, raise serious doubts about its identity over time.

The second point is the same as that which I made about deterministic laws. Even if one believed that there were ideal situations in which mechanical systems such as (some) machines were totally deterministic, in the real world machines interact with all kinds of things the deterministic quality of which is much less plausible. Cars encounter floods and sandstorms or human vandals with sugary fuel additives that counteract any tendency to deterministic behavior. And thus even if one believes that the blueprints in the Toyota design division represent a perfectly deterministic system *ceteris paribus,* in the outside world other things are never equal. And once again determinism if anywhere must be everywhere.

I now turn to the second claim with which this section opened, that determinism is contradicted by the majority of our most successful scientific theories. The obvious starting point for this discussion is quantum mechanics, which is often taken to be the most fundamental theory we currently possess (although I am skeptical of the relevant sense of "fundamental" here) and also to be irreducibly probabilistic. Since I am unqualified to assess the highly abstruse and technical arguments that are currently discussed in connection with this claim of irreducible indeterminism, I shall leave this issue as an untapped reserve of possible support. What I do want to stress is the prevalence of theories that explain apparently quite deterministic behavior in terms of the interactions of very large numbers of constituents homogeneous in certain relevant respects. The classic case here is thermodynamics, seen as deriving from the statistical mechanics of vast numbers of molecules. Similarly, if the indeterministic nature of quantum mechanics is assumed, electrical phenomena should be seen as involving the indeterministic behavior of enormous numbers of electrons. Macroscopic chemical reactions must be understood similarly, and indeed are even strictly stated in terms of stable equilibria between the backward and forward directions of the reaction.

The same strategy, though with less of an appearance of generating deterministic laws, can be seen in parts of the biological and social sciences. In biology the prime examples are selective processes in evolution. Sufficiently strong selective pressures acting on large populations could generate probabilities of a certain kind of evolutionary change so high as to approximate determinism. Yet it is well understood that the relevance of this selective force to particular individuals can be anything but deterministic. Similar claims are made in economics with regard to such things as the effects of changes in price on consumer demand, although perhaps here the appearance of a deterministic structure is becoming too weak to take seriously.

So far I have been considering cases in which something approximating deterministic behavior is grounded in an underlying microindeterminism. In all such cases we can say that if this is the correct grounding, the laws in question cannot be strictly deterministic. Even in thermodynamics, it is generally acknowledged as possible, if almost infinitely unlikely, that, for example, all the gas in a container should come together in the middle of the container, reducing the pressure on the walls to zero. And it is certainly possible that evolution will sometimes produce a less fit organism simply by chance.

In addition to examples of this kind, there are many others in which nothing approximating to deterministic behavior emerges at any level. Such are the theories of domains such as meteorology or ecology that I mentioned in arguing for the lack of empirical grounding for determinism. It should be stressed once again that if it is conceded that *any* of these domains are truly indeterministic, the determinism of all other domains is threatened by interactions.

Let me now approach the issue from a quite different direction and consider the question how, at a very general level, a doctrine of determinism might be articulated. All such articulations of which I am aware take the basically Humean tack of attempting to characterize the universal laws that ground the deterministic causes linking particular events. With this much in common, two rather different approaches to specifying the relevant laws can be discerned. The problem on which I want to focus is that of identifying the laws underlying everyday causal relations ("The rise in interest rates caused the recession," "Massive trauma to the head caused his death," "Smoking caused her lung cancer," and so on). The first of these approaches, already encountered in Chapter 7, is the idea that the universal laws must be sought in terms quite distinct from the ontological level of the commonsense causal relata. The same idea is deployed, though in very different versions, by Davidson and by the Churchlands. Since all versions of this approach presuppose the kind of reductionism that we have seen good reasons to reject, I shall not consider this strategy any further here.

The other approach, developed by Mill (1875) and refined to a high degree of sophistication by J. L. Mackie (1974), attempts to generate universal causal laws by sufficiently detailed elaboration of the relevant causal factors at the structural level at which the commonsense causal connection lies. On Mackie's analysis something is a cause of a certain effect if it is an insufficient but nonredundant part of an unnecessary but sufficient condition (or, acronymically, an *inus* condition) for the occurrence of that effect. Among competing inus conditions we often denominate something as *the* cause on pragmatic grounds. Thus we would say, for example, that the death of civilians in an air-raid shelter during bombardment was caused by the bombing of the air-raid shelter rather than by their presence in the air-raid shelter. If the air-raid shelter had been a military installation, we might rather attribute their death to their presence at such a location. Both putative causes are inus conditions. But since it is to be expected that civilians will be in an air-raid shelter during an air raid, the latter is treated as a background condition and the bombing as the cause. Mackie's definition specifies that the cause will be insufficient in recognition of the inevitable presence of such additional background conditions in formulating anything like a sufficient condition. The sufficient condition is said to be unnecessary in recognition of the fact that (most) effects can be brought about in more than one way.

For present purposes the crucial part of this analysis is the sufficient condition, embodying the assumption that there is some law of nature dictating that whenever that set of circumstances obtains, an effect of precisely that kind will occur. Mackie's analysis provides a brilliant account of the structure of our commonsense causal knowledge. But any attempt to push such an analysis to the extreme of offering a recipe for constructing universal laws of nature is quite implausible. (So far, indeed, I agree with skeptics about commonsense causal relations such as the Churchlands.) This implausibility is rendered most apparent by the observation that the purportedly sufficient set of conditions will, as Mackie is well aware, have to include not only the constellation of conditions together capable of producing the effect, but also the absence of any conditions that could intervene to prevent the effect. (This requirement echoes the problems with interactions faced by supposedly determinist theoretical laws.) Certainly no one is likely to spend much time trying to construct such a list of possible intervening conditions; and it seems highly likely that the limits of such a list are merely the limits of human imagination and perseverance. More important perhaps is the following: even if we were to succeed in specifying a Mackian sufficient condition, the great list of auxiliary conditions and absent preventers shows us just how rapidly the causal nexus begins to diverge as we trace causal chains or processes back beyond the immediate antecedents of an event of interest. And this divergence shows just how pressing is the problem to which I keep returning, that of interaction with indeterministic causal chains. Any of the list of background conditions or absent preventers could be produced or generated by some causal chain with indeterministic links. To provide an adequate grounding for a deterministic universe we need to be sure that *every* causal connection is underwritten by a completely sufficient condition. That this can be done in terms of the concepts at the structural level in question is incredible, so that the retreat to reductionistic accounts of determinism is easy to understand. But reductionism is false. So, therefore, is determinism.

I shall conclude by considering briefly a kind of example that figures largely in discussions of this topic. Gambling devices such as dice, tossed coins, and roulette wheels are often cited as illustrating our experience of chance. But such examples are also widely used to convey the message that chance is really only an epistemological condition, whereas the world may very well be deterministic. Thus it is often said that a tossed coin has a probability of ½ of landing heads, in the sense that we have no way of assigning a higher probability to one outcome or the other until the coin finally lands. But, it is also said, once the coin is tossed, the initial conditions of the toss will completely determine the remainder of the trajectory. It is often supposed that this is a paradigm for apparently chance events, and thus that chance is merely a reflection of our ignorance. As Hume put it: "Though there be no such thing as *Chance* in the world; our ignorance of the real cause of any event has the same influence on the understanding, and begets a like species of belief or opinion" (1748, p. 37).

The weakness of this position should by now be clear. First, it is just false that the way the coin will land is determined by the conditions of its launch. If it encounters massive interference—collision with a passing swallow, say, or a rifle bullet—its trajectory will surely be influenced. Less dramatically, it is quite plausible that normal random motions of molecules in the air might be sufficient to affect the final outcome. But the most fundamental problem is the lack of evidence for what should seem nothing more than a powerful metaphysical prejudice. Why should such a process be assumed to be deterministic at all? Determinism, as I indicated at the beginning of this chapter, is just one of a range of possible metaphysical positions on the prevalence of causal order. We should not assume it without either strong empirical evidence or a compelling (and empirically supported) theoretical grounding. I see neither in such cases. The prima facie empirical evidence is that this is an indeterministic process, with probabilities of ½ for each of the two possible outcomes. I see nothing that overrides the claims of this prima facie empirical evidence to provide the most reasonable belief.

I conclude, then, that we must descend down the scale of causal order at least as far as probabilistic uniformitarianism. As a matter of fact, probabilistic uniformitarianism is susceptible to problems analogous to those I have just been considering. But whereas determinism is seldom defended in much detail now, even if it is widely assumed by writers in the philosophy of mind or moral psychology, versions of probabilistic uniformitarianism are currently widely debated and defended. In the next chapter, in opposition to most of this discussion, I shall try to show that probabilistic uniformitarianism inevitably collapses into catastrophism. And this conclusion, finally, leaves it an open empirical question how far the world is from true randomness.

9. Probabilistic Causality

The Standard Account and Its Problems

Having established in the preceding chapter that we have no reason to believe in any more causally ordered world than one characterized by probabilistic uniformitarianism, in this chapter I shall consider in detail some recent ideas about probabilistic causality. I shall argue that the uniformitarian aspect of these theories is to be seen as largely deriving from residual deterministic thinking. As with determinism itself, the assumption of uniformity may be methodologically useful, but as a belief about the real causal order of the world it is unsupportable.

As I have already shown, probabilistic causality encounters difficulties in its application to particular cases. In the first instance, at least, it is a relation between *types* of events, states, or processes. And roughly, the relation in question is that a cause should raise the probability of its effect; that is, the occurrence of an instance of the cause type should increase the probability that an instance of the effect type will occur.[1] The major problems that arise in providing a more detailed and more nearly adequate analysis of probabilistic causality even as it relates to event-types derive from the fact that the particular cases in which these types are instantiated will certainly differ from one another in various respects. And frequently they will differ in respects that are causally relevant to the outcome. Thus, to recall the stock example, smoking may raise the probability of lung cancer, but there is no number x such that the chance of one's getting lung cancer given that one smokes is x%; for the value of x will surely depend on the amount one smokes, the way one smokes, whether one has also been exposed to asbestos or polluted air, and much else besides.

In fact there are two causally significant modes of variation that should be distinguished. The first, which I have just illustrated, consists of the different causally relevant contexts in which the cause may occur. In addition, the cause may affect the probability of its effect in more than one way, most obviously through the production of intermediate effects with differing or even opposite influences on the probability of the final effect. Different routes may link the cause to the effect. The first of these problems has been widely recognized in formulations of probabilistic theories of causation, and an orthodox response has emerged. The second has been less widely discussed. I shall address both here. I shall begin by explaining why differing contexts have been seen as presenting a problem, and outlining the solution that has been offered. I shall argue that the solution is unacceptable, and that the correct conclusion to draw from this problem is serious skepticism about the uniformitarianism that underlies this solution.

The most pressing problem posed by varying contexts has come to be known as Simpson's paradox. Consider, to vary slightly the antismoking message so prominent in this literature, the supposition that smoking causes heart disease, but that smoking is also highly correlated with a tendency to exercise regularly, perhaps because both the disposition to exercise and the disposition to smoke are themselves caused by some unknown genetic factor. Assuming that exercising discourages heart attacks, it is possible that this benefit might outweigh the effect of smoking to such an extent that smokers actually had fewer heart attacks than non-smokers. It might still be true that smoking causes heart attacks. This could be demonstrated by dividing the population between those who do, and those who do not, exercise regularly, and by discovering that in each of these groups smokers were more liable to heart attacks. This kind of case is a major part of the reason for the truth of the familiar observation that correlation does not prove causation.

A similar example has been discussed by Cartwright (1983, p. 37). It was once observed that female applicants for graduate programs at the University of California at Berkeley were much less likely to be admitted than male applicants. This finding immediately suggests the hypothesis that being female causes one to be rejected by graduate schools. Surprisingly, further investigation suggested that this was not the case. When applicants were sorted according to the departments to which they had applied, it turned out that in most departments the acceptance rates were about equal, and in some they were even higher for women. The explanation of the discrepancy was just that women applied disproportionately to departments with low overall admission rates.[2]

Such cases motivate what I have referred to as the orthodox solution to the problem of different contexts. The solution is a generalization of the procedure suggested in the smoking and heart disease case. It proposes that a factor is a cause if and only if it raises the probability of its effect in *every possible* context of other factors relevant to the production of the effect. As this view is generally expressed, we must partition the data with respect to all possible combinations of additional causally relevant factors, and determine against the background of each such combination that the cause does indeed raise the probability of the effect. Brian Skyrms (1980) allows a slightly weaker condition, that in some context the cause must raise the probability of the effect, and in no context must it lower that probability. If there is some context in which the probability of the effect is lowered, then, according to Ellery Eells and Elliott Sober (1983), "there will be no such thing as *the* causal role of smoking with respect to heart attacks in the population as a whole" (p. 37).

This last claim is argued in more detail by Cartwright (1979). She observes that if a cause fails to raise the probability of its effect there must be a reason. Either the reason is a correlation of the kind illustrated in the smoking and heart disease case, or the cause should be treated as interacting with some feature of the environment. Two causes are interactive if in combination "they act like a causal factor whose effects are different from at least one of the two acting separately" (pp. 427–428). Hence the interactive cause should be treated as a distinct case from the cause in noninteractive cases, and presumably Cartwright would agree with the conclusion just quoted from Eells and Sober. I shall call this condition on probabilistic causes the requirement of contextual unanimity.

The same example may be used to illustrate the second problem, the problem of different causal routes. Suppose we raise the question whether the hypothetical genetic factor responsible for both smoking and exercising is a cause of heart attacks. The problem is that the genetic factor has two effects, one of which tends to cause heart attacks, and one of which tends to prevent them. This question has been discussed at some length by Eells and Sober (1983). They pose the question as one about the transitivity of causation: given that the gene causes smoking and that smoking causes heart attacks, does it follow that the gene causes heart attacks? First they note that a sufficient condition for such transitivity is that all the paths from cause to effect have a positive effect on the probability of the cause. They refer to this condition as the condition of unanimity. Transitivity is threatened only if there is some route from the cause that decreases the probability of the effect. To distinguish it from contextual unanimity I shall refer to it as unanimity of intermediaries. Eells and Sober deny that this condition is necessary; all that is necessary is that the influence of those intermediaries that increase the probability of the effect should outweigh the influence of those that decrease the probability. Thus in the example, the question whether the genetic factor causes heart attacks turns only on whether the probability of having a heart attack is increased or decreased by the presence of the genetic factor (given, presumably, that some route from the genetic factor to the increased likelihood of heart attacks constitutes a series of steps satisfying the condition of contextual unanimity). In their discussion of contextual unanimity Eells and Sober remark that "average effect is a sorry excuse for a causal concept" (p. 54). It is curious that in the case of differing intermediaries this is exactly what they propose.

It would be very difficult to apply anything analogous to the condition of contextual unanimity to the case of intermediaries. In the example we are considering the genetic factor was hypothesized to affect both the probability of smoking and that of exercising. There is no reason to suppose that it may not increase both these probabilities in all, or just the same, contexts. Thus there is no relevant way of partitioning the contexts in which the gene occurs to limit the causal claim to only those cases in which smoking is in fact made more probable.

But if averaging is inevitable in response to this problem, why should we not accept it in the case of differing contexts? An obvious response is that we want to avoid making causal claims based on spurious correlations, as in the case of the smokers with a reduced incidence of heart attacks. But although this is undoubtedly a desideratum, it does not entail that we must go so far as to define causality in terms of wholly homogeneous partitions. And I want to claim that there are overwhelming reasons for rejecting this move.[3]

My first point is that it seems highly doubtful whether the acceptance of averaging over intermediaries is even consistent with the insistence on contextual unanimity. Suppose that statistics for the United States showed that possession of the relevant gene in fact increases one's chance of having a heart attack; in other words, the effect of the increased tendency to smoke outweighs the effect of the greater propensity to exercise. Then the account by Eells and Sober indicates that the gene does indeed cause heart attacks. But then suppose that we were to collect statistics in some part of the world where tobacco was unavailable. Presumably we would find there that the effect of exercising would predominate, and thus that the gene was a preventer of heart attacks. This transforms the issues into one of different contexts. The availability of tobacco is surely a contextual factor causally relevant to the occurrence of heart attacks, and thus it appears that the gene does not exhibit contextual unanimity in causing heart attacks. And so, according to the orthodox account, it does not cause them.

This problem will surely be quite general. Whenever there are different routes by which a factor may influence its supposed effect, the relative significance of these different routes is likely to depend on aspects of the context that facilitate or impede the various possible causal paths. Thus it is very likely that there will be different contexts in which each type of effect predominates. And to the extent that this is so, the insistence on contextual unanimity will force us to accept unanimity of intermediaries also as a necessary condition of causal connection.

The point can be clarified by considering the example that Eells and Sober use to illustrate their views. They discuss two phenomena widely recognized as pervasive in genetic systems and often taken to present difficulties for the thesis of genic selectionism or, more specifically, for the claim that the presence of a single gene may cause the reproductive success of an organism. Polygenic effects, in which the effect of a particular gene depends crucially on the genetic context, are seen as presenting a fundamental obstacle in the light of the requirement of contextual unanimity; pleiotropy, on the other hand, in which a gene has more than one effect on the organism, turns out not to be a problem in view of their position on causal transitivity. But certainly the relative importance of the various pleiotropic effects of a gene will depend on features of the environment, both at the organismic and the genetic levels, so that pleiotropy will be just as much of a problem as polygeny for the genic selectionist.

The solution to this difficulty that I advocate is to abandon the requirement of contextual unanimity. In support of this proposal let me offer one more elaboration of the main example. Suppose that scientists employed by the tobacco industry were to discover a rare physiological condition the beneficiaries of which were actually less liable to heart disease if they smoked than if they did not. Contrary to what the orthodox analysis implies, I do not think that they would thereby have discovered that smoking did not cause heart disease. Had this condition been the rule rather than the exception, we would rightly have concluded that smoking was a prophylactic against heart disease. And neither would this conclusion have been refuted by the discovery that those abnormal and unfortunate individuals who lacked this physiological advantage were actually more susceptible to heart disease if they smoked. If these intuitions are correct, they suggest that causes should be assessed in terms of average effect not only across different causal routes, but also across varying causal contexts. In support of this construal of the example I note, finally, that it is almost inconceivable that we should ever exclude, conclusively, the possibility of such a fortunate minority. And thus insistence on contextual unanimity really does drive us to agree with the tobacco companies that there is no evidence that smoking causes disease.

Eells (1987) has responded to the arguments just expounded.[4] His main point is that my arguments ignore a crucial aspect of the theory of probabilistic causality as it has been developed by recent writers. This is that it is a relation not just between a cause and a probabilistic effect, but between a cause, an effect, and a *particular population*. Thus what we should say in the preceding example is that smoking causes heart disease in the population of those individuals who lack the physiological anomaly, whereas in the population that possesses the anomaly, smoking prevents heart disease. For the total population it is a "mixed" factor, neither causing nor preventing heart disease. Similarly in the earlier example, smoking causes cancer in the U.S. population but prevents it in the population of the area without tobacco. But although I do not for a moment deny that the discovery of this anomalous condition would provide a significant supplementation of our knowledge of the effect of smoking on heart disease—a decisive supplementation for smokers who should happen to discover that they had this condition—I deny that it contradicts our previous belief that, for the population in general, smoking really does cause heart disease.

The immediate problem with Eells's response is that the epistemological embarrassment mentioned at the end of my introduction of the physiological anomaly example is simply transferred from the question whether the alleged cause really is a cause of the effect in question, to the equally intractable question, What is the population in which this cause can be said to act? Although I agree that probabilistic causal claims must generally be understood as applying to specific populations,[5] I take these to be much more commonsensical populations than the potentially highly esoteric, and as yet unknown, populations required for Eells's theory. If "Smoking causes heart disease" turns out not to be true for humankind, I take it that its truth has been established for the various populations of Western countries in which it has been systematically investigated. It is now known, for example, that smoking causes heart disease for the human population of the United States. There may be inductive hazards in the extrapolation of this result to other identifiable human populations; but it seems unnecessarily cautious to restrict the claim to an as-yet-unidentified subpopulation of inhabitants of the United States.

This is only the beginning of the epistemological troubles of the contextual unanimity thesis. Suppose there are n factors causally relevant to the production of an effect. Then unanimity requires a positive or neutral effect in each of the 2^n combinations of these factors. For many cases of interest it seems likely that 2^n will exceed, or be of comparable magnitude to, the population in question. The human population, for example, certainly among the largest we are likely to deal with in biology (and often much too broad a category for generalizations about humans), is of the order of 2^{31}. It seems perfectly plausible that the probability of heart attacks might be affected by as many as 31 factors. If so, the average size of the causally homogeneous classes would be about one; many would be empty; and it would be quite impossible to determine whether the unanimity condition obtained. It might well turn out, for a causal factor with complex and variable interactions with other factors, that the population in which that factor was operative could be specified only by complete enumeration—although even then, I have no idea how it might be identified.[6]

It is always possible to hide from such epistemological difficulties behind a metaphysical figleaf. Eells (1987) suggests that some of my early criticisms of the unanimity thesis were misdirected because motivated by merely methodological considerations; whereas the unanimity thesis was intended rather as an account of what probabilistic causality really consists in. I shall have a good deal to say about the metaphysics of probabilistic causality later in this chapter. For now I shall propose only what I take to be a highly plausible, if for some tastes excessively empiricist, principle of metaphysical methodology: We should avoid metaphysical doctrines for which we neither have, nor possibly could have, empirical evidence of applicability. This is a methodological principle that the unanimity thesis fails dismally to satisfy.

An Alternative Account of Probabilistic Causality

The starting point for a more adequate conception of probabilistic causality is the realization that the unanimity thesis is driven by a quite inappropriate paradigm of scientific investigation. More specifically, the goal of finding a complete set of circumstances that are both individually necessary and jointly sufficient for an effect, a goal that in principle at least is mandated by the commitment to determinism, is naturally translated into the complete partitioning requirement we have been considering in the probabilistic case. Both embody the ideal of a complete causal story, a story that records the role of every factor, positive or negative, present or absent. But given, at least, the nonexistence of *sufficient* conditions in a genuinely probabilistic case, a major part of the ideal of causal completeness must surely go. The arguments of the preceding section suggest that we should also dispense with much of the rest. The ghost of causal completeness, not fully laid by the rejection of determinism, is liable to prevent us from pursuing more appropriate approaches to probabilistic causality. In particular, an investigative methodology that provides a more promising starting point for an analysis of probabilistic causality is the use of controlled experiments.

This standard methodology involves drawing two samples at random from the population under investigation. To the members of one sample, the test sample, is applied the factor under investigation. The prevalence of the effect in question in the test sample and in the other, control, sample is then compared.[7] A statistically significant difference provides evidence that the putative cause does indeed tend to produce the effect. To indicate how far removed this idea is from what I have been calling the orthodox account, it should be noted that for someone who believes that the unanimity condition is an essential part of an account of (probabilistic) causation, the importance attaching to controlled experiments in science is quite mysterious. For a controlled experiment tells us only about average effects, and nothing about the various positive and negative effects that may contribute to that average. Consequently they can never provide us with grounds for believing that the experimental treatment causes the effect in question in every possible causal context, and for a proponent of the unanimaty thesis they cannot, therefore, provide any information at all about causality.

A decisive test of whether smoking causes heart disease, then, would be to take a large sample of human infants randomly selected from the human population, divide them into two equal groups, and force one group to smoke for the rest of their—no doubt abbreviated—lives.[8] The cases that have been mainly debated by myself, Eells, and others are those in which such an approach is contingently unavailable—in the present case, for obvious ethical reasons. The point of a controlled experiment, of course, is to neutralize the distortion due to interfering factors by averaging their effects. This, it seems to me, is the best methodology we have for investigating probabilistic causes. A plausible empiricist principle closely related to the one mentioned at the end of the preceding section would be that we should assume no more about the world than is necessary to account for our most successful investigations of it.[9] I take it that to assume more order, in the sense loosely defined by the spectrum of positions introduced at the beginning of the last chapter, is to make a stronger assumption than to assume less order. The argument that we are inclined to make unwarrantedly strong assumptions about the prevalence of such metaphysical order is a central theme of this book. The application of such a principle suggests the following proposal for defining probabilistic causation: if we compare otherwise fair samples in one of which the cause is uniformly present and in the other the cause is uniformly absent, the effect will occur with greater frequency in the sample in which the cause is present.

As Eells (1987) points out, the notion of "fair sample" in this account needs some work. Even in cases in which a controlled experiment is possible, there may be significant difficulties in finding a method that will select a random sample of the population being investigated. But in cases such as the smoking/heart disease question and many similar investigations of naturally occurring populations, there can be no question of drawing a random sample from the population, for the simple reason that we want a sample of *smokers*. The problem, then, is that by simply selecting a (random) sample of smokers we will also select a sample containing any other factors correlated, for whatever reason, with smoking. The best possible recipe for generating a fair sample in such a case, I suggest, would be to select samples of smokers and nonsmokers from the population in such a way that the frequency of other known causal influences on the effect under investigation, in this case the occurrence of heart attacks, is the same in these samples as the frequency with which they occur in the general population.[10] We conclude that smoking (probabilistically) causes heart attacks just in case the frequency of heart attacks is higher in the sample of smokers than in the sample of nonsmokers. The underlying rationale is just the same as that for a controlled experiment: we aim to average out the influence of other, perhaps unknown, causal factors by comparing samples in which other causes occur with equal frequency—equal, in fact, to the natural population frequency. In the latter case we have a method with a good chance of bringing this about; in the former we can do no better than try to cook up samples with this feature.

The method I am advocating cannot be guaranteed to produce correct results, simply because the recipe refers only to known causal influences. It is always possible that unknown influences will provide misleading results, since our samples may differ with respect to the frequency of the unknown factor. The best possible attempt at providing fair samples may yet miss its mark. Similarly, of course, the existence of unknown causal factors will defeat the attempt to provide an exhaustive partition. No one has a monopoly on fallibility. However, it should be clear that my proposal brings method and metaphysics a good deal closer together than can the unanimity theory: whereas both methods can take account of all (and only) the factors known to be relevant, the metaphysical account in terms of (objectively) fair samples provides a much less stringent account of the cause we are looking for than does the account that incorporates the unanimity condition. It is this weakening that gives my account the great advantage of allowing that controlled experiments can, indeed, deliver causal knowledge.

The Metaphysics of Probabilistic Causality

The unanimity thesis implies that there is some finest level of description under which we can specify the complete causal truth about a situation. Perhaps there is no fact of the matter whether an individual event *was the cause* of a succeeding event; but at least that event can be shown to have had some quantitatively precise positive or negative effect on the probability of its successor. Absent some minor concerns about causal overdetermination, it is easy to see how the assumption of complete causal truth fits a

deterministic world. The complete causal truth consists in sufficient and generally necessary conditions' linking one state of affairs
with the next. The deeper point that I claim follows from the
rejection of the unanimity thesis is that the assumption of a complete causal story must be rejected with the determinism that gave
rise to it. The basic ground for this claim is quite simple: causal
completeness depends on the presence—indeed omnipresence—of
universal laws, laws that specify in complete detail everything that
can affect the occurrence of the effect. Mackie (1974) has given
the canonical account of how this should be done in a deterministic
world; the requirement of contextual unanimity does the same
thing for the probabilistic case. Either approach pursues the Humean premise that to understand a cause is to identify the regularity in accordance with which effect follows cause. What I want
to argue is that if we take probabilistic causality seriously—not,
that is to say, as merely reflecting our ignorance of the real deterministic processes—this Humean thesis must be abandoned. In
fact we must completely invert this picture. What are fundamental
are just what Hume rejected: the causal capacities of objects or
events.[11] These are revealed in the statistical regularities that occur
in the populations in which those individuals exist, and it is those
statistical regularities that provide the evidence for the capacities.
One of the reasons why we should not feel compelled to seek the
full specification of the laws governing causal processes is that if
this picture is correct, there is no reason to believe that there are
any such laws.

A good place to pursue my skepticism about quantitatively
precise causal regularities is with a further point of disagreement
between myself and Ellery Eells. My example of the two populations in which tobacco is and is not available, and of the respec-
tive predominance of each of the different effects of the genetic
factor hypothesized to cause both exercising and smoking, suggests the rather surprising conclusion that the laws of nature might
turn out to depend upon contingent facts about the populations
to which they apply. If, say, a virus were completely to wipe out
every tobacco plant on Earth and no substitute were made available, then the genetic factor that had previously been a cause of
heart disease would have become a preventer of heart disease.

Thus the truth of a law will depend on the obtaining of a particular fact. (This is true, incidentally, both on my account and on Eells and Sober's [1983] account of causal transitivity.) I say "surprising" because if one thinks that laws are among the fundamental facts we hope to discover about the universe, such dependency will seem a serious embarrassment. However, I do not see why the conclusion need be disturbing. Since Hume, most philosophers have come to accept that the truth of laws is contingent. If the laws are probabilistic, then what happens is contingent even given the laws. It strikes me as a small step to accept that the laws themselves may be dependent on what happens.

Eells does not find this conclusion either necessary or acceptable, but his response strikes me as remarkable. For he appears to concede that insofar as the laws in question apply only to specific populations, my claim is true. But he then dismisses the suggestion as being true in only "a trivial sense" (1987, p. 112). Citing Hempel's (1965) distinction between fundamental and derived laws, he asserts that my claim applies only to the latter. But earlier in the same paper Eells is at pains to assert that probabilistic causality is (really, metaphysically, is) "a relation between *three* things: a causal factor C, a probabilistic effect of it E, and *a population within which C is a causal factor for E*" (p. 107; emphasis in original). It follows immediately, since all probabilistic laws apply only to specific populations, that all probabilistic laws are merely derivative, that there are no real probabilistic laws as opposed to mere statistical facts about particular populations. My suspicion that the unanimity thesis conceals a kind of covert determinism could hardly have received more striking support.

I imagine that most theorists of probabilistic causality, including Eells, would not wish to conceive of *all* probabilistic laws as necessarily relativized to particular populations. The laws of quantum mechanics, at least, are not generally conceived in this way. But Eells and I perhaps agree that the laws of biology, anthropology, economics, sociology, and so on—that is, the vast majority of science—must be so relativized. It appears that Eells wants to claim that these laws are all derivative, and therefore scarcely worthy of being denominated "laws."[12] And of course to claim that the laws of most of science are thus merely derivative is to make a strong commitment to reductionism. Once again we see major pieces of the mechanistic world picture fitting tightly together.

It is easy to see the way out of this difficulty if we abandon the Humean primacy of laws in favor of treating causal capacities as fundamental. The problem is that the genetic factor we are considering has two capacities. It is true both that it has the capacity to cause heart attacks (by encouraging smoking) and that it has the capacity to prevent heart attacks (by causing exercising). In populations similar to our own such a gene would presumably exercise *both* these capacities, and its net effect on heart attacks would be the resultant of both effects. Given that there are interacting factors that can favor the exercise of one capacity over the other, such as the availability or perhaps even price of tobacco, this net effect will depend on the distribution of such other factors in the population in question. One might, perhaps, on this picture, say that the laws were derivative, not from other more "fundamental" laws, but from the causal capacities, or causal powers, of the individuals that make up the population. However, since I do not take capacities to be necessarily mathematically quantifiable, the laws derivable in this way will not satisfy the traditional picture of laws of nature.

The point of view I am advocating also gives a rather different perspective on the methodological issues concerning the goals of partitioning in the investigation of causal hypotheses. What Simpson's paradox shows is that statistical investigation may suggest the presence of a causal capacity where there is none. A population characterized by the gene we have been considering might display a high correlation between smoking and resistance to heart attacks. But smoking does not have the capacity to prevent heart attacks. We investigate causal interactions, and control for other factors as best we can, to avoid being misled by such cases. But the goal, I suggest, is not to converge on the complete, true law, but to get the most reliable evidence as to whether smoking does indeed have the capacity to cause heart disease, and to estimate as well as possible the strength of that capacity in a population of concern to us. If this account even makes sense, there is no reason to assume that any such complete, true law exists; or that there is any finite list of factors that have the capacity to cause or prevent heart attacks, or to interact with other factors that have this capacity. From a methodological point of view again, it may well be asked how we know when to stop looking for further causes. The answer is that we cannot *know*. The question when we have adequate evidence for the existence of a causal capacity (in a case in which appropriate controlled experiments are not possible) is one of those problems that make the doing of science an art rather than the mechanical procedure sometimes suggested by philosophical analyses.[13]

The possibility of dual but opposed capacities is of great significance to the position I am advocating.[14] First, as the discussion of different causal intermediaries should have begun to suggest, this is by no means an esoteric possibility.[15] In complex domains I suspect it is the rule rather than the exception. In evolutionary biology, for example, almost any trait short of the lethal or drastically physiologically deleterious is likely to have possibilities for both positive and negative contributions to fitness depending on features of the environment. (The imaginative endeavors of "Panglossian" biologists are sufficient to have established at least that much.)

But if the prevalence of dual and opposed capacities is admitted, a striking consequence immediately follows. Wherever there are dual capacities, any attempt to characterize the causal situation in general terms is bound to conceal a significant part of the causal facts, simply because there are indefinitely many combinations of opposed capacities that will produce the same frequency of an effect in a population under investigation. There might be just the capacity to produce the effect, or a much larger capacity to produce the effect combined with a significant capacity to prevent it. Clearly the strength of these capacities can vary indefinitely provided only the difference between them remains constant. We cannot hope to separate the population into subpopulations in which only one of the capacities can be exercised, because unless we are covert determinists there is no reason why both capacities might not have the potential to act in the very same cases.[16] This difficulty confirms my agreement with the claim of Eells and Sober (1983) that causal analysis requires averaging over intermediaries. But it also reinforces the importance of my argument that averaging over intermediaries implies the inevitability of averaging over contexts. And the necessity of averaging over opposed capacities shows that the facts about capacities cannot be reduced in the Humean manner to facts about regularities.

A further consequence of the conception of probabilistic regularities as reflecting the exercise of capacities of members of populations, is to reinforce a central theme of the earlier parts of this discussion, that there is no reason to expect that indefinite investigation will lead us to probabilistic laws converging on some true and ultimate measures of the strength of a cause. I would like to illustrate this point with a rather more extended example from an area that, though hardly central to science, is perhaps the most prolific source of statistics in American society, the game of baseball.

Analysis of the determinants of the outcomes of baseball games does not, in practice, appeal to canonical abstract accounts of baseball players. Rather, we define types of events as precisely as possible and treat the relevant interactors (baseball players, umpires) at many different levels of abstraction. The heterogeneity of these interactors is explicitly admitted but largely unproblematic. The rules of baseball provide a conceptual structure that allows us to sort events and investigate various statistical patterns among those events. In certain respects this case can be seen as providing an ideal paradigm of statistical investigation of a complex domain of interacting entities, so that the example will help to elucidate by analogy and contrast the general limitations on this kind of investigation. I note three main respects in which baseball provides an ideal object of statistical investigation. First, baseball games are unproblematically teleological. There is only one significant result to a baseball game: one team wins and the other loses. Second, this teleological aspect defines unequivocally which events are relevant to the outcome of interest. The only way of winning games is to score runs, and the only way to score runs is to advance around the bases to home plate. Of course, there are a number of ways to move around the bases, but these are pretty much exhaustively specifiable. And we know that such matters as how often the hitter spits are completely irrelevant except insofar as they may affect these primarily relevant events (perhaps by disgusting the pitcher and thereby disturbing his concentration). And third, these events can be infallibly detected and unambiguously assigned to the appropriate categories. Although the criteria for these events admit of occasional interpretative problems, all actual ambiguity is removed by giving definitive authority to make such determinations to particular participants, umpires. One scores a run, for example, if and only if an umpire adjudges one to be safe at home plate. (Baseball would no doubt lose some popularity if these decisions were not perceived as being generally in accord with more objective criteria.)

The existence of this clear and unambiguous structure facilitates a variety of investigations of the causal determinants of the outcomes of baseball games. Most familiar are the widely distributed statistics about the frequency with which particular participants achieve particular desirable results (getting hits, reaching base, driving in runs, stealing bases, and so on). More analytic students of baseball investigate more-general and more-specific questions. More-general questions are, for example: What is the general frequency with which players get hits? How has this changed over time, and why? What is the typical tradeoff between the probability of getting a hit and the probability of hitting home runs? How effective are different styles of pitching, and how do they effect the likelihood of serious damage to the pitcher's arm? More specifically one may investigate the different characteristics of particular players facing left-handed pitchers or right-handed pitchers, hitting during the day or at night, and so on. Many of the most interesting questions prove quite difficult to answer, especially those relating to the efficacy of particular strategies. When, if ever, does attempting to steal a base or to execute a sacrifice bunt increase the probability that your team will win? How should players with different skills be most effectively ordered in a batting lineup? These are complex questions to investigate, although it is fairly clear what kind of information is relevant.[17]

Apart from providing a paradigm of a structure that determines what questions are worth asking and gives fairly clear guidelines about how they can be answered, this example is intended to make a more fundamental point, again about causal completeness. Consider a specific interaction, a particular batter facing a particular pitcher in a particular game situation. There are many useful generalizations that are relevant to such an interaction. The hitter will have some established record of competence (lifetime batting average, on-base percentage, slugging percentage), as will the pitcher (earned-run average, strikeout-to-walk ratio, and so on). Perhaps the batter has a significantly different record of competence against left-handed pitchers or at night. Perhaps these very individuals have faced each other sufficiently often to provide useful information. The crucial point is that there are invariably enough potentially relevant factors to reduce any such interaction to sheer particularity. Even when the very same players have faced each other a substantial number of times, they have not done so in this park, at night, after an off-day, while the hitter is in a slump, with the pitcher perhaps just passing his prime and having had a fight with his lover that morning, and so on. The question what level of specificity is relevant to assess the particular probabilities of salient outcomes is a difficult matter of judgment, and a kind of judgment that baseball managers are very well paid to exercise. And of course this judgment is not thereby arbitrary. It would certainly be possible, to pick up another fashionable heuristic, for a careful student of these matters to make money betting at suitable odds on the outcomes of these interactions.

The general point is to note the modest metaphysical status of the kind of general knowledge that is relevant to an ideal knowledge of the determinants of baseball games. Baseball statistics can be indicative of quite genuine and stable *capacities* of baseball players. That Tony Gwynn will have a higher batting average this year than Jose Uribe is about as likely as that emeralds will continue to be green. Such a prediction is grounded in differing capacities: Gwynn, but not Uribe, has the capacity, for example, to hit .320 over a season. But though our knowledge of such capacities must be based on statistical facts (for example, Gwynn's career batting average higher than .320), we must recognize that these statistical facts are not only a part of many more abstract statistical patterns, but also subsume many more-detailed patterns. And as we move from the smaller-scale patterns to specific events there is no finest-grained general pattern to be found. We do not aspire to the complete causal story; we move from general knowledge to the specificity of historical narrative.

It is easy to see that the problem of opposed capacities applies readily to this case. Thus, for instance, a particular pitcher may have a disposition to prevent the hitter from getting a hit, say by throwing a perfect curve, and also the capacity to cause a hit, say by hanging a slider. Different pitchers may have the same probability of giving up a hit by combining different blends of these two capacities.[18] Thus the most complete general account of the statistics, fully contextualized to every set of additional causal factors, and even supposing baseball players to live forever to eliminate the problem of vanishing reference classes, would still fail to give a fully determinate account of the underlying causal properties.

Might there nevertheless be a God's-eye story in terms of the instantaneous physiological state of the pitcher and the hitter, the air density and movements between the two, and so on? One suspects, for a start, that the sensitivity of the contact between a round stick and a round ball to details of relative position might be sufficient to allow quantum indeterminacies in the action of the atmosphere on the flight of the ball to have significant effects. So such an account might not, even given the most solidly physicalistic assumptions, be deterministic. But be that as it may, the important point is that such a story plays no role in either our initial understanding or our more sophisticated investigation of events of this kind. And so, finally, to the extent that this is an appropriate paradigm for macroscopic causal interactions, it provides no basis for a belief in causal completeness.

If the outlines of the preceding analysis are correct, it should be clear that statistical investigation of natural phenomena, lacking the three features mentioned above, will be all the more problematic. Consider first teleology. With no idea what a thing is doing, or aiming at, it is quite unclear what causal capacities we should be investigating. Various sciences have partial solutions to this problem. In biology we often assume fitness as an implicit goal, although the dangers of excessive reliance on this assumption are well known. In economics goals such as profit maximization or, at the macrolevel, growth are generally assumed. Here the dangers are not merely epistemological. The assumption of such goals is liable to slide into the covert assumption of the appropriateness of these goals, and to foster the concealment and promotion of such side effects as systematic exploitation and environmental destruction. In reality, what *should be* the goals of economic systems is one of the most fundamental problems of the discipline. Perhaps, to turn to the third point, it is reasonably clear what events contribute to these purported economic goals. In biology, by contrast, what features contribute to survival and reproduction, and how, are extremely vexing questions. That there exist laws determining the evolution of biological systems, in which the parameters are variables that can be discovered by statistical investigation, is even less likely than that there is an algorithm determining even the exact objective probability for each team in baseball to win the World Series.

A world in which the degree of causal order that exists is attributable only to the presence of causal capacities associated to a more or less determinate degree with certain properties or kinds is, then, a world in which we have no reason to expect the orderliness of probabilistic uniformitarianism. This is an expectation that we should abandon with the unanimity thesis. Certainly there will be frequencies with which capacities are exercised in particular populations. (That is a matter of logic.) Since those populations will generally be subject to change of various kinds over time, we should expect these frequencies to change. Insofar as there are *any* laws governing the impact, say, of smoking on the incidence of heart disease in the human population, we should expect that such laws would include parameters that varied unpredictably over time. Devotees of causal order are likely to insist that the interesting question is rather whether, if we could somehow hold fixed every other factor causally relevant to the occurrence of human heart disease, the effect of smoking would then be constant. The short answer, which, it seems to me, is the right one, is that we cannot hold all such factors fixed, so we cannot possibly know what would happen if we did. This impossibility reflects not just experimental difficulties, but also the fact that the human population is constantly changing. More generally, I suggest that the quantitative stability of causal capacities is entirely an empirical matter. A reasonable expectation would be that some were quite stable, and others perhaps hardly stable at all.[19]

I imagine that many people would, if convinced that there was indeed no persuasive evidence for a quantitatively complete causal nexus at the macroscopic level, conclude that this was all the more reason to retreat to the microlevel and provide a complete causal story in terms of less recalcitrant entities. That is, they would fall back on reductionism. I have said enough about my general assessment of reductionism. I would like to make one point about this retreat to the microlevel in the present context. Our intuitions about causality, it seems to me, are derived wholly from the macrolevel. This is particularly clear by contrast with the strange discussions of causality characteristic of contemporary microphysics. But even if we turn to entities rather larger and better behaved than electrons or quarks, such as DNA molecules or neurons, even if we assume determinism for methodological reasons, this assumption is not based on experience of the fully explained behavior of individuals of these kinds. If experience rather than methodology grounds such an assumption anywhere, it is for macroscopic objects such as machines or organisms. *Belief in*—as opposed to merely the methodologically motivated assumption of—causal completeness is transferred from the macroscopic to the microscopic rather than vice versa. So reductionism, even if it is not rejected on independent grounds, should not threaten my conclusions here.

A Note on Freedom of the Will

Although the problem of freedom of the will is generally understood as involving a conflict between human freedom and determinism, the rejection of determinism has not generally been perceived as providing a solution to the problem of human freedom. There are good reasons for caution here. It has occasionally been suggested that quantum mechanics, by showing that certain events can occur without determining antecedent conditions, shows how human actions could be free. Perhaps a movement is somehow generated by an amplified quantum event. But a solution to the problem of freedom of the will is not to be obtained by replacing the picture of a person as mindless machinery with that of a random action generator. The idea that people act for reasons seems more reconcilable with an account under which those reasons turn out to be nothing but states of the machine, than one which seems to preclude either causes or reasons by placing the action wholly beyond explanation. Nor is it helpful to move from determinism to a probabilistic uniformitarianism that claims that human actions are not determined by antecedent physical conditions, but only made more or less probable. This sounds not so much like an account of a (metaphysically) free person, as of a somewhat unreliable one.

But the effect on our conception of human freedom of relaxing the assumption of determinism can be seen quite differently. One of the problems with the quantum amplifier account of freedom is that, just as with human freedom under traditional libertarian accounts, the quantum indeterminacy is treated as a source of anomaly in a world otherwise driven by inexorable laws. A more radical rejection of complete and pervasive causal order suggests a quite different but much more promising approach.

To the philosophically uncorrupted, what is striking about humans as compared with other natural phenomena is not their unpredictability or their role as sources of disorder in an otherwise reliable and well-ordered universe. Quite the contrary: it is people who appear as somewhat orderly and predictable entities in a generally chaotic world. Where we find order, it is generally because it has been intentionally created by other humans. When I make an arrangement to meet a colleague for lunch sometime next week, I become moderately confident that she will be at a particular determinate place at, say, 12.30 P.M. next Wednesday. It is inconceivable that I could make a similar prediction about the scrap of paper that blows by as I am making the arrangement— unless I pick it up and decide to bring it along. The point is just that humans have all kinds of causal capacities that nothing else in our world has: capacities most notably to make and execute all manner of plans capable of determining their behavior in complexly organized ways for considerable distances into the future.[20] There is no good reason for projecting these uniquely human capacities in a reductionist style onto inanimate bits of matter. Nor is there anything ultimately mysterious about particular causal capacities' being exhibited uniquely by certain very complex entities; no more than it is insolubly mysterious that certain weather patterns can generate tornadoes capable of picking up houses or that leguminous plants have the capacity to fix nitrogen. Thus the solution I propose to the free will problem is simply to recognize that what is problematic about humans is not that they are exceptions to an otherwise seamless web of causal connection, but that they are extraordinarily dense concentrations of causal capacity in a world in which such order is in short supply. Scientific investigation of the sources of these capacities may help to explain how this is possible; but this is to begin to explain, not, as determinists often want to suggest, increasingly to problematize, this remarkable feature of humans.

It is natural, on the view I am presenting, to feel considerable sympathy for one aspect of Kant's account of freedom in terms of autonomy (1785/1948). It is the ability to do what one thinks one ought to do rather than what, at that moment, one would like to do that is, arguably, uniquely human, and that makes possible the execution of long complex schemes. Unlike Kant, I do not mean to restrict what one ought to do to a narrow sense of duty (although conceivably this may serve to distinguish a qualitatively distinct and higher form of conduct), but to include any prudential "ought" driven by any kind of human project. Thus planning an armed robbery or a racist political advertisement is, in the sense I am considering, as much a display of uniquely human capacities as building a hospital or dismantling a nuclear weapons system.

The most relevant differentiation from Kant's position that I want to insist on is that without the commitment to determinism or even causal completeness, I am under no pressure to engage in the kind of metaphysical excesses in the noumenal world that have made Kant's position so implausible. I deny that there is any problem in reconciling the possibility of directed, intentional human action with the surrounding causal order. As I argued in Chapter 6 in connection with evolutionary change at the organismic versus the genetic level, there is no reason why changes at one level may not be explained in terms of causal processes at a higher, that is, more complex, level. In the case of human action, the physical changes involved in and resulting from a particular action may perfectly well be explained in terms of the capacity of the agent to perform an action of that kind. Unless there is some reason to think that some part of the physical surroundings is undergoing some distinct and contrary causal process, there need be no conflict.

A final point is relevant. One reason for some thinkers' insistence on freedom of the will is their belief in the uniqueness not only of humans as a kind,[21] but also of human individuals. How far it is *true* that individual humans are, in the intended sense, unique is a question beyond the scope of this book. But certainly the position I have outlined makes it entirely possible. The bundles of capacities instantiated by particular humans will differ substantially from one to the next; and, in my view, none will be swept along inexorably by the tide of exceptionless laws. I have denied earlier that humans will be all equally constrained by common possession of any human essence. Thus the extent of human individual uniqueness will be a metaphysically open question, and must be determined by more detailed and specific investigations.

I do not suppose that this brief discussion can contribute substantively to the elucidation of the nature and variety of human nature. But the metaphysical point of view that I have been developing throughout the book can help to eliminate some of the most notorious obstacles to such elucidation. I shall also take up some of the claims about the broad metaphysics of human action in connection with the human sciences at various points in the final chapters.

IV. Some Consequences of Disorder

10. The Disunity of Science

The conceptions of an ordered nature and a unified science belong naturally together. If there is some ultimate and unique order underlying the apparent diversity and disorder of nature, then the point of science should be to tell the one story that expresses this order. The reductionist version of this picture considered in Part II of this book has contemporary atomistic materialism as the ultimate truth, and physics as the science that aims to express this truth. If, as I have argued throughout the preceding chapters, we reject the assumption of *any* such systematic and universal underlying truth, not only should we give up the specific version expressed by contemporary materialism or physicalism, but the motivation for any unified account of science becomes questionable. However, the unity of science has been, or might be, defended in a number of ways quite different from the reductionist interpretation most generally associated with this phrase. Here I shall look at a number of important such interpretations and argue that no thesis of the unity of science can serve any legitimate purposes for which it might be intended. Consideration and criticism of these ideas will also enable me to elaborate and deepen the motivation for my advocacy of scientific disunity.

Sociological Unity, or the Unity of Scientism

I begin with the kind of scientific unity that I think has the greatest plausibility. Unfortunately, it is a kind of unity that has no genuine consequences either for metaphysics or for epistemology. It is the unity that follows merely from the fact that the prestige of science in contemporary Western societies is such that "scientific" has become an epistemic honorific quite independent of any general consensus about what makes scientific claims any more deserving of credit than beliefs from any other source. The entitlement to this honorific derives, rather, solely from the institutional status of the persons from whom the claims originate. The extent of this hegemony is unintentionally parodied in the absurd or banal claims made by actors in white lab coats in television advertisements. The dangers of this pseudoepistemic power are depressingly illustrated by such spectacles as the parade of professional "scientists" available to testify on behalf of what George Bush once referred to, as aptly as ironically, as "voodoo economics."

It is this merely sociological unity, the source of the epistemic authority that accrues to anyone simply by being institutionally certified as a scientist, that I refer to facetiously as the unity of scientism. (On this conception science and scientism, of course, may have a substantial overlap.) Insofar as the concerns of this book *are* metaphysical and epistemological, this kind of unity is irrelevant to my skepticism about scientific unity. The contrast between scientific and scientistic unities does, however, point toward a vital epistemological problem, one that was central to the philosophical concerns of Karl Popper under the rubric of the "demarcation problem." This is simply the problem of what distinguishes a science, in the sense of a body of opinion that deserves epistemic authority, from a pseudoscience, or something that has only the institutional trappings of a science. I do not mean this formulation to imply that *only* a science can have epistemic merit. It may well be more difficult to distinguish legitimate scientific from nonscientific knowledge production than it is to distinguish serious from spurious scientific projects. It is the second such distinction with which I am concerned, and which I refer to as the "demarcation problem." Indeed, I remain agnostic whether the first can be drawn at all. Perhaps in part because Popper's solution to the demarcation problem in terms of falsifiability has been almost universally rejected, this problem has ceased to be a central concern of philosophy of science. In my view this is extremely unfortunate; indeed I can see nothing more urgent to the agenda of that discipline. I do not, I regret to say, have a detailed solution to that problem to offer. I shall, however, at the end of this chapter suggest a pluralistic approach to the question of genuine scientific merit.

One other aspect of what I mean by "scientism" has some bearing on matters addressed in earlier chapters. Whereas a degree of scientificity is attained merely from the fact that opinions originate from an academic or other institution staffed by individuals with doctoral degrees in some discipline generally counted as part of science, there are formal aspects of the products of such institutions that tend to amplify greatly their degree of epistemic prestige. Most obvious of such aspects is the enormously enhanced respect that accrues to parts of science that give a central role to mathematical methods. It is not at all my aim to banish mathematics from science. Mathematics has a wide range of uses in various scientific enterprises, many of them entirely legitimate. The probabilistic views of causality discussed in the preceding chapter, for example, reinforce the importance of the mathematical analyses of statistical data. And large parts of physics and physical chemistry appear to have proved amenable to systematization in terms of mathematically often quite precise laws. But the extension of this approach to many areas in which no such empirical successes can be claimed is part of what I refer to abusively as scientistic.

That this aspect of scientism—perhaps we should call it "mathematicism"—is a sociologically significant contributor to scientific prestige seems hard to dispute. It is again perhaps best illustrated by the preeminent influence of economics, with its characteristic appeal to abstruse mathematical models of little empirical worth, among the social sciences. A number of factors probably contribute to this situation. Most relevant to my general theme, more or less inchoate commitments to reductionism surely play some part in the continuing role of physics, and indirectly its mathematical methods, as the paradigm to the methods of which other sciences should aspire.[1] More broadly still, believers in an ultimately orderly universe often suppose that mathematics provides the language necessary to capture such metaphysical order. Straightforwardly sociological factors surely also contribute to mathematicism. It seems very plausible, for example, that mathematical complexities the mastery of which requires substantial training, by setting up barriers to entry (to borrow a concept, this time, from economics), serve to increase the financial and other rewards that accrue to membership of scientific professions. Or even more simply, a mystifying veneer of mathematics will sometimes serve to conceal the banality of what is offered as scientific wisdom. At any rate, the omnipresent neo-Pythagoreanism of contemporary science is surely not adequately justified by its empirical successes. If it is motivated by any legitimately theoretical considerations, I suspect that these amount to some kind of commitment to a universe amenable to one systematic and orderly description; a universe in the existence of which, I have argued, we have every reason to disbelieve.

Whatever one's level of skepticism about various scientific enterprises, it should be clear that sociological unity provides a problem rather than a rationale for the unity of science. In the absence of some account of scientific unity strong enough to legitimate the whole range of practices that are generally counted as scientific, it seems extremely likely that some very questionable candidates may sneak in under the cloak of sociological unity. Sociological unity raises a pressing need, therefore, for some theoretically compelling solution to the demarcation problem, a solution that legitimates the epistemic pretensions of the various disciplines that count as science by justifying a uniform treatment of scientific belief.[2] Such a solution would establish a substantive unity coextensive with the sociological domain of science. In what follows I shall consider some possible candidates for this role. If, as I suspect, none proves adequate, individual disciplines must be left to sink or swim on their specific merits. It seems to me very likely that, if such a conclusion were truly taken to heart, substantial parts, at least, of a number of disciplines would sink without trace.

Scientific Unification versus Scientific Unity

Not everything that has been discussed under the rubric of scientific unity or scientific unification is relevant to the problem just raised. A quite disparate set of issues has on occasion been raised in connection with scientific unity.[3] In particular, a number of philosophers have discussed processes of scientific *unification*, processes that form links of various kinds between parts of science or extend the range of application of scientific theories or techniques. However, the identification of such processes need have little bearing on the questions about overall scientific unity that are my present concern. To clarify what kind of scientific unity would be required to create a presumption of credibility for arbitrary parts of the sociological whole, it will be helpful to look at some recent philosophical appeals to scientific unification that cannot be expected to fill such a role.

One account of scientific unification, explicitly opposed to any kind of traditional reductionism, has been presented by Lindley Darden and Nancy Maull (Darden and Maull, 1977; Maull, 1977). Darden and Maull distance themselves from the reductionist tradition by criticizing the assumption that areas of scientific investigation should be identified with particular *theories*. Instead, they propose a more complex and generally more local classification of scientific domains, into scientific *fields*. A scientific field is to be identified by a number of elements: a central problem, a domain of facts related to that problem, explanatory goals and expectations, appropriate techniques and methods, and sometimes concepts, laws, and theories thought relevant to the realization of the explanatory goals (Darden and Maull, 1977, p. 44).[4] Examples they discuss include cytology, in which the central problem is the characterization of different kinds of cells and their functions; and genetics, for which the central problem is the explanation of patterns of inheritance.

Various relations may exist between fields. One field may aim to explain the nature or structure of entities postulated in another field, or there may be causal relations between entities postulated in different fields (Darden and Maull, 1977, p. 49). Such relations may pose problems the solution of which calls for what they call *interfield* theories. An example is the chromosome theory of Mendelian heredity, as a theory lying between the fields of cytology and genetics. These interconnections, finally, suggest a possible reconception of the unity of science. When we move from the classical reductionist's hierarchy of scientific theories to Darden and Maull's more realistic classification of science in terms of fields, we note that there are no sharp boundaries between these parts. Indeed many important theories lie precisely in the interstices between such fields. Thus, Darden and Maull hypothesize, "unity in science is a complex network of relationships between fields effected by interfield theories" (p. 61).

I agree with most of this. Certainly some much more fine-grained classification of scientific domains, more closely related to real scientific practice, strikes me as a great improvement on the broad categories assumed by classical reductionism. And I have no objection to the idea that there may be close connections of the kind Darden and Maull describe between such fields. The only point I want to insist on, possibly though not certainly in disagreement with Darden and Maull,[5] is that the consequences of these proposals do nothing to encourage any kind of global thesis of the unity of science.

First, as Darden and Maull explicitly state (1977, p. 45, n. 5), their conception does nothing to address the demarcation problem, to distinguish science from nonscience or pseudoscience. So it does not say what fields might, in their sense, be unified. Their ideas do, perhaps, suggest a bolder thesis. Might the continued development of links between theories eventually provide a densely connected network such that all and only the fields that truly belonged to science were connected to this network? Exploration of this possibility will begin to reveal some of the difficulties with even such a weak conception of scientific unity.

To begin with, it seems entirely possible that the extrapolation of this process of connection of fields by interfield theories would produce not one connected whole, but a number of such wholes. Clearly, the possibility of such a bridge between cytology and genetics or even—to take the more traditionally hierarchical example described by Maull (1977) as generating interlevel theories—between genetics and biochemistry does not show that there might be a theory linking, say, electronics and cultural anthropology; or even that there might be a sequence of theories leading from one member of this pair to the other.

But even supposing that there were, in the ideal future of science, enough interfield theories to provide a sequence of links connecting any two scientific fields, this would not be sufficient to establish, in any interesting sense, the unity of science. For it is hard to imagine that the conception of an interfield theory capable of realizing this possibility could be sufficiently restrictive to exclude connections reaching out to areas that we would not normally think of as part of science. Thus, for example, there is a natural area of theoretical investigation lying between linguistics and literary theory, the latter of which is not generally considered to be part of science. Or, for that matter, there are many theories that lie between fields of philosophy and of science. The Ghiselin/Hull theory of species as individuals, for example, might reasonably be said to lie between metaphysics and systematics. Many philosophers would no doubt welcome the discovery that philosophy was just an area of science. But the conclusion of this whole line of thought now seems to be just that no form of knowledge production can be entirely isolated from all others. This seems far too banal an observation to glorify with the title "unity of science"; and certainly it does nothing to legitimate the parts of such a unified science.

The problem might be put as follows. Darden and Maull successfully identify a kind of connection that can be forged between different parts of science. But their characterization of these connections presupposes that the two fields being connected are, indeed, part of science. In the absence of a solution to the demarcation problem, this does nothing to illuminate the question how many such connections would be needed to form a unified science. Only a characterization of interfield links with the consequence that all and only parts of science are connected can unify science. And this, finally, will be possible only if science is antecedently a unified whole, which is, of course, the question at issue.

A quite different appeal to the unification of science, again explicitly opposed to reductionism, is to be found in the work of Philip Kitcher (1981). Kitcher's discussion starts with the question, deriving from well-known difficulties with positivist accounts of explanation, What makes an argument to a certain sentence expressing some fact an explanation of that fact? The answer he develops is, very roughly, that arguments are accepted when they instantiate argument patterns that, in turn, are accepted because of their capacity to generate large numbers of conclusions that we take to be true. Kitcher sees scientific progress, at least in the sense of the provision of increasingly better explanations, as consisting in the production of theories with ever-wider ranges of explanatory potential. If progress involves movement toward theories of ever-broader range of application, it would thereby seem to involve a move in the direction of scientific unification. And indeed Kitcher does claim explicitly that unification should be taken to be a criterion of scientific explanation.

It is not relevant to this discussion to address the important issues about explanation that are the main targets of Kitcher's account. For however convincing is the case for the thesis that, at a *local* level, extension of the range of application of a kind of explanation is crucial in achieving acceptance for scientific innovations, it implies nothing about the probability that science will eventually undergo *global* unification. Progress (if such there be) in, say, biology, could go on indefinitely instantiated by a sequence of theories able to explain ever more *biological* facts, but would do so by achieving ever-more-powerful understandings of the same domain of phenomena. Kitcher's argument implies that *if* a theory were introduced that explained in approximately the same way, for example, a large range of both chemical and biological phenomena, it would be accepted over a theory that explained only the same biological phenomena. But there is no reason to believe that such theories are generally available, and much reason to doubt it.

The situation is parallel to that for another desideratum often proposed for scientific theories, simplicity. If this is indeed a desideratum, then if two theories have equal merit in terms of every other relevant characteristic, but one is simpler than the other, we should accept the simpler theory. But this principle tells us nothing about how simple or complex the best theory may ultimately turn out to be. If the world is very complex, then presumably simple theories will fail dismally in other respects, such as empirical adequacy. Thus, similarly, I have no objection to unification as a desideratum of scientific theorizing. But how much unification is consistent with otherwise adequate theorizing is another matter.

A general moral can be drawn from the discussion of the (quite different) ideas of Kitcher and of Darden and Maull. Unification may be significant for science in many ways. But an account of the significance of unification tells us nothing about the existence of unity. Treaties may exist, and be a wonderful thing. But these facts do not imply that there is a world government.

Methodological Unity

Perhaps the commonest view of science as unified is grounded in the assumption that there is a distinctively scientific *method*. Although this idea attributes no unity to the content of science, it does suggest the possibility of a criterion satisfied by all and only those beliefs, or perhaps practices aimed at the production of beliefs, that belong to the domain of science. What people most often have in mind when they praise an opinion as "scientific," I suppose, is that it was generated according to the correct canons of scientific method, though no doubt the criterion for application of this epithet is very often no more than the institutional source of its subject (see my comments above on sociological unity).

Surprisingly, however, methodological accounts of what it is to be scientific have little currency in the contemporary philosophical literature. This is not to say that philosophers have no interest in methodology, although there is less attention devoted to such issues than to, say, questions about the form or structure of scientific theories. It is rather that philosophical concerns with methodology are generally very local, either addressed to specific areas of science or, more commonly, concerned with quite particular and often technical aspects of such matters as, for example, the significance and techniques of randomization in experimental design. One reason for this rather narrow focus is undoubtedly the recent attention directed by historians of science to the details of scientific experiments rather than merely to the theories these experiments were supposed to support, attention that has revealed the remarkable particularity of the actual practice of science in specific research programs.[6] Few, if any, philosophers are currently concerned with conceptions of methodology of the kind that might provide general criteria of scientificity.

The twentieth-century philosopher who *did* attempt to provide a methodological criterion for science was, not coincidentally, the philosopher most centrally concerned with the issue of demarcation, Karl Popper (for example, 1959, 1963). As is well known, Popper's idea was that science was distinguished by the commitment of scientists to attempt to falsify, by experiment or observation, their professional beliefs. This, of course, is one attempt to make concrete the generally inchoate assumption that what is distinctive about science is its adherence to the tribunal of empirical evidence. Popper's falsificationist criterion of scientificity has, however, been almost universally rejected. The grounds of this rejection suggest one very general difficulty with a methodologically grounded account of the unity of science.

The idea of falsificationism is that the basic function of a scientist is to test theories. This is to be done by inferring empirical consequences of scientific theories and discovering whether, in the real world, those consequences obtain. If they do not, the theory is refuted. If, on the other hand, a theory survives many such attempts on its credibility, it becomes increasingly worthy of being believed. It is perhaps a virtue of falsificationism that it does not admit the possibility that a theory could be *verified* or proved true. This issue has generated a great deal of discussion, but it is not relevant to my present concerns. Additionally and notoriously, it provides no account of the source of scientific ideas. The etiology of scientific ideas is taken to be a matter for psychology rather than for scientific methodology (hence the much-discussed distinction between the context of discovery and the context of justification). However, the problem that has led to the widespread rejection of the theory, although it has considerable bearing on the latter issue, is neither of these.

The central difficulty for falsificationism comes at the very beginning, with the process of inferring empirical conclusions from a theory and attempting to falsify them. The problem is that such inferences will typically require vastly more by way of premises than merely the theory that is supposedly being exposed to possible falsification. But then it is quite unclear what can be concluded from a negative result to an experiment. *Perhaps* the theory is wrong, but perhaps some other of the many auxiliary premises is the culprit. Given the likelihood that scientists will have strong commitments to the theories they devote their professional lives to investigating, it seems entirely reasonable that they should seek out any other possible explanation of an apparently anomalous result before rejecting their theories. But if that is true, and if they will almost always have alternative possibilities to investigate, it is hard to find any plausible role for the goal of falsification in actual scientific practice.

Consideration of more-concrete examples readily illustrates the extent of this difficulty. Even in what might seem a relatively simple case, such as the observation of planetary positions as a test of Newtonian dynamics, many assumptions go into making a prediction that could be exposed to the risk of falsification: an accurate knowledge of the positions at some earlier time, a full knowledge of the existence and positions of all relevant celestial bodies, the absence of other intervening forces, the ability to make sufficiently accurate observations of the predicted position and the correct functioning of any instruments, such as telescopes, that may be used in generating such observations, and so on. It is most unlikely that there will not be some among these assumptions that are more readily given up than the fundamental theories underlying research in that area of science. A classic case is that of the inference by Leverrier and Adams from unexplained irregularities in the orbit of Uranus to the existence of the planet Neptune.[7] These irregularities were not, needless to say, taken as refuting Newtonian mechanics, but rather as motivating a search for the erroneous assumption, a search that culminated in the prediction of the existence and orbit of Uranus. More telling still, Leverrier also noticed deviations in the orbit of Mercury. But here no such explanation was found prior to the replacement of Newtonian with relativistic mechanics. Thus this latter case proved in the end to constitute, by our current lights, a genuine falsification of Newtonian mechanics. But no one took it as such until a new theory that actually predicted the phenomena in question had been developed as a serious rival to the Newtonian account. If, instead of these cases from astronomy, one thinks rather of the kind of observations generated by the vast and elaborate machines of contemporary particle physics, and contemplates merely the number of components that must be assumed to function in a determinate way, the prospect of falsification becomes completely hopeless.

This problem is fundamental to contemporary epistemology and philosophy of science. Since the difficulty was recognized, in essence, by Duhem and is closely parallel to the epistemological holism made famous by Quine (1960), the idea that it is generally impossible for a particular piece of evidence to bear exclusively on a unique proposition for which it is to be taken as evidence for or against has come to be known as the Quine-Duhem thesis. It is an essential motivation of Imre Lakatos' (1978) radical reconstruction of Popper's views into his own "methodology of scientific research programmes," a theory that allows only for the gradual refutation of entire projects of inquiry. It can also be seen as a crucial part of the background to Kuhn's (1970) theory of scientific revolutions. What these last two theorists share, though strongly opposed in many other ways, is the understanding that routine scientific research, at least, is carried out taking fundamental theoretical beliefs for granted. Such beliefs are not subject to simple empirical refutation and, indeed, typically coexist with well-known prima facie empirical refutations. It is this recognition that gives rise to the still pressing problem of what, if anything, could make fundamental scientific change rational, a problem to which Lakatos and Kuhn develop very different and very influential answers.

It should now be possible to see why these developments have tended to undermine the project of discovering the unity of science in a uniform and distinctive methodology. One might reasonably hope that an account of methodology would illuminate the day-to-day practice of science, what Kuhn has called "normal science." But if the most basic of scientific beliefs are not called into question in such quotidian scientific activity, then methodological considerations will apparently account neither for these beliefs nor, perhaps more importantly, for changes that occur in them. It seems clear, on the contrary, that to explain why biologists believe in the theory of evolution, or why geology is guided by the theory of tectonic plates, we need a historical account of the process by which consensus on these theories was reached. Of course, it is conceivable that such a historical account could be told in terms of the successive applications of some homogeneous scientific methodology. But a vast body of work in the history of science, especially over the last thirty years, argues against such a hypothesis. As Kuhn began to convince the philosophical world with *The Structure of Scientific Revolutions,* major scientific changes are enormously more complex than this. Indeed since Kuhn's work, though opposed to some extent by Kuhn himself, there has been an increasing tendency to downplay the role of *all* systematic, reason-driven factors in the explanation of scientific change and to focus increasingly on broader ideological, political, and other extrascientific factors. But whether or not this tendency has been pushed too far, there remains no plausibility at all to the prospect of an account of the history of science in which progress is constantly generated by application of a uniform methodology. This fact explains my final observation on the topic, that since the demise of Popper's falsificationism there have been no serious rivals for the role of a universal methodological criterion of scientificity.

Science as a Process

The preceding discussion suggests a rather different tack that has seemed much more promising to contemporary philosophers of science. Accepting that the history of science cannot be unified by the discovery of some thread of transcendent methodology leading inexorably toward the truth, there might still be features of the practice or institutions of science capable, in the long run, of selecting those ideas that are nearest to the truth. Thus, returning after all to something close to the sociological unity with which this chapter began, scientific unity might be discovered as a feature common to the institutions and social processes of science, but a feature that really did legitimate the epistemic authority enjoyed by those institutions. Something of this sort has had currency for some time under the rubric of evolutionary epistemology. The general idea is that knowledge advances as different beliefs compete one with another, those being selected that prove most beneficial to their proponents. (In many contexts, it is piously hoped, truth will tend to make a belief more beneficial.) Starting from this conception, if there is some way of setting up the institutions of knowledge production so that beliefs are somehow selected exclusively for truth, empirical adequacy, or some such suitably exalted epistemological virtue, then it might be reasonable to call just those institutions set up in this way the sources of genuinely scientific belief. Thus could a more charitable interpretation be put on what I described dismissively as the sociological unity of science. The most detailed proposal along these general lines has been developed by David Hull (1988), and this will be my focus for the remainder of this section.

The first point that is essential to an understanding of Hull's thesis is his treatment of scientific ideas as historical individuals.[8] The relevant sense of "scientific ideas" is a broad one that includes concepts, terms for theoretical entities, more or less theoretical beliefs, and experimental or other investigative techniques (Hull, 1988, p. 434). Hull's claim is that scientific ideas should be considered as forming lineages, passed from scientists to their colleagues and students.

Scientific ideas are promulgated and promoted by individual scientists and by scientific research groups, and properties of these will affect the success of the idea lineages that they adopt. Research groups are also seen as subject to selection processes, and their success will determine the future of the ideas that they adopt. Presumably the ideas adopted will have some reciprocal influence on the success of the groups and individuals who adopt them. The more one inclines to a rationalist, progressivist view of science, the more importance one will be likely to accord to this reciprocal influence. The functioning of the tightly connected groups of scientists that Hull has in mind in discussing research groups is illustrated, as is the remainder of Hull's thesis, by an extraordinarily detailed and fine-grained history of recent systematics.

Hull's general thesis can be expressed as follows in the useful terminology that he himself favors. Scientific ideas, again broadly construed, are replicators. Research groups, individual scientists, and perhaps other entities are interactors, and their behavior is such as to affect differentially the propagation of the replicators. Thus processes of selection act directly on scientists and research groups; those that prosper—increase in size, train more graduate students, and so on—will thereby succeed in propagating the ideas that they champion. I shall return shortly to Hull's account of how this selection progress leads to epistemological progress. But first I shall say something more about his account of scientific ideas.

The conception of scientific ideas as lineages analogous to biological lineages is in many respects highly illuminating. For one thing it immediately defuses a range of traditional concerns about conceptual change. Rather than focus on the adoption by scientists of a sequence of, by our lights, false beliefs, Hull's picture encourages us to focus on the causal relations between sets of beliefs held by scientists at different times. Rather than ask whether scientists at one time used terms with the same reference as their predecessors, to which the answer, with boring predictability, is no, we can instead focus on whether particular ideas are parts of the same conceptual lineages, and to what extent they resemble or differ from their ancestors. And finally, the picture of competing conceptual lineages provides a much richer taxonomy of conceptual change than, for example, the sharp Kuhnian dichotomy between normal science and scientific revolutions. We can conceive of gradual evolution within a lineage, something like normal science; competition between closely related lineages in which some thrive and others go extinct; or, finally, colonization of an existing scientific niche by a more distantly related lineage, an event analogous to a Kuhnian revolution.

My reservations about Hull's proposal run closely parallel to my reservations (explained in Chapter 2) about Hull's treatment of species as individuals. My enthusiasm for the proposal rests on my conviction that Hull has provided us with an illuminating perspective on science. I suspect, however, that his claim would be that he has provided us with the *correct* perspective on science. And as in the case of species, I doubt whether there is any such thing. In some contexts it is desirable to treat species as individuals, in others as modest kinds. The same goes for ideas. Whereas the treatment of scientific ideas as lineages can be very helpful when applied to the development of highly technical concepts such as "monophyly" or "electron" as these have been employed in iden-

tifiable scientific traditions, its appropriateness becomes questionable for more abstract terms—terms, perhaps, such as *interactor*
and *replicator*. I have a number of reasons for wanting to insist on
such a distinction, but here I shall focus on what I take to be the
most important, that a distinction must be maintained between a
descriptive, and a critical or normative, account of science.[9] If we
are interested in explaining the development of a set of scientific
ideas over time, then tracing lineages of belief transmission may
be exactly the right thing to do. On the other hand, the function
of Hull's distinction between interactors and replicators, it seems
to me, is the inculcation of conceptual clarity. Of course, such an
essay in conceptual clarification may or may not affect the relevant
lineages of scientific belief. But despite Hull's insistence that scientific ideas that fail to replicate are irrelevant, I would like to
suggest that the value of his conceptual analysis is less of a hostage
to fortune than this suggests. A conceptual lineage can be concep-
tually confused and demonstrably so even if none of its constituent
interactors, or only a marginal and impotent few, ever know it. I
do not deny that persuasion may in some sense be a lot more
important than rightness. But we should beware of obfuscating
the distinction between the two.

The last example concerns epistemological norms. More important even than these, the distinction between descriptive and
normative accounts of science is necessary for the possibility of
articulating externalist normative critiques of science. Thus not
only do I doubt whether the selective processes Hull describes will
necessarily lead toward greater objective truth, but I fear that they
may foster opinions that are positively deleterious to human well-
being. But consideration of these kinds of normative issues will
be postponed to the final chapter.

Let me now return to the details of Hull's evolutionary account
of scientific progress, in particular to the way in which he claims
it is apt to generate progress in our understanding. Hull offers an
intriguing and detailed hypothesis as to how the individual goals
of scientists serve to generate a process at the social level that
seems so strikingly successful in advancing our understanding of
many phenomena. Scientists, he suggests, are motivated by curiosity, by credit, and (negatively) by checking. Curiosity, he pro-

poses, is the prerequisite for anyone's getting involved in science in the first place. Once involved in science, scientists seek credit for their contributions. Their pursuit of credit is constrained by the expectation that their more ambitious claims are likely to be checked by others, and exposure of error, or worse, of deceit, is both likely and very costly. Because science is essentially a cooperative and cumulative activity in which one practitioner builds on the work of others, scientists must give credit—typically in the form of citations—as support for the premises on which they build. The more useful a scientist's results are to others, therefore, the more credit she or he will receive. Scientists working in close collaboration or as common defenders of a particular theory or research strategy will naturally cite one another's work more frequently. Such collaboration will tend to increase both the credit of themselves and their co-workers and the credibility of the project in which they are all involved. It also explains in part why Hull thinks integrated research groups will generally be more efficacious interactors than isolated individuals. In the long run, Hull argues, this process will lead to selection of the most-valuable scientific ideas, or replicators.

Hull draws on his rich and detailed account of recent systematics for empirical support of these ideas. Although much of this is not especially edifying—politicking, struggles for control of journals, obstructing publication of unsympathetic research, and all manner of skullduggery—Hull argues that what we find is ultimately a highly productive process of competition among different research groups for credit and thereby the replication of their views. As well as the details of these wars of words and deeds between pheneticists, cladists, and evolutionary taxonomists, Hull provides detailed analysis of who cited whom, what was published where and with what obstacles, and so on. At the very least Hull's hypothesis has earned the careful consideration of all students of scientific change. Here, however, I shall not attempt a direct evaluation of Hull's views, but rather shall indicate a difficulty that arises immediately given my focus on scientific unity: What is to count as a science to which this account is intended to be relevant? Hull seems to think that what he is offering is a quite general account of how science progresses. What

I want to suggest, on the contrary, is that its relevance will vary greatly from one prima facie scientific field to the next. This suggestion will also serve to elaborate my opinion that even at the level of sociological process, sciences are quite disparate activities.

There are two reasons why a satisfactory answer to the question of the extension of science is of crucial importance to the evaluation of Hull's thesis. First, Hull is suggesting a general mechanism for *science*. If, as is my inclination, one thinks of science as consisting of a loose and heterogeneous collection of more or less successful investigative practices, then it will be implausible to suppose that any such general mechanism exists. Moreover, in assimilating science to the class of selection processes, of which the paradigm is presumably natural selection, Hull is clearly proposing a *science* of science. Since it is not obvious that science is the kind of thing of which there could be a science, the evaluation of Hull's proposal requires that we have some prior conception of what a science is, and what kinds of phenomena can provide the subject matter for one. One of the assumptions that the present discussion aims to problematize is that something similar to a paradigmatic science is the appropriate way to study any arbitrarily selected range of phenomena.

A simple answer to this demarcation question would be to suggest that the mechanism that Hull describes should be a *criterion* for science. Thus one might suppose that science was a natural kind, and that an essential feature of the kind was that it exemplified the kind of sociological processes that Hull describes. But it is pretty clear that this feature, even if necessary, is not sufficient. Not only marginal sciences such as history, but hardcore humanities such as philosophy and literary criticism appear to exhibit the curiosity, credit, and even checking that drive Hull's proposed mechanism of scientific change. Certainly philosophers cite authorities in support of their claims, and reject the claims of other philosophers as inadequately argued, empirically false, incoherent, and so on. Moreover, it is easy enough to distinguish relatively tight groupings of philosophers pursuing particular programs and characterized by such things as high levels of mutual positive citation. Whether, on the other hand, the existence of such a process leads inexorably to progress is, to say the least, controversial.

Further differentiae seem required, and indeed a number of additional criteria seem plausible. First, which Hull mentions several times in his book, there should be some essential connection with dependence on empirical evidence. Second, again emphasized by Hull, we expect science to be cooperative and cumulative. This is indeed implied by Hull's proposed mechanism. Third, we might expect a genuine science to include, or at least seek, laws or principles of considerable generality. I shall now briefly consider whether it is plausible that any set of such conditions might be jointly sufficient or individually necessary for a human practice to count as a science. My negative answer will point to the diversity of investigative practices that I am currently urging.

The first condition, though in principle perhaps the least controversial, is notoriously difficult to formulate adequately. Relations between theory and evidence tend to become increasingly obscure the more interesting and powerful the theories in question. Nevertheless, there is little doubt that many of our most cherished scientific theories, at least in the physical sciences, exhibit a striking coherence with empirical data. The issue becomes much murkier with those areas of science that attempt to provide theoretical models of intelligible simplicity for admittedly enormously complex processes. I have in mind, for example, population genetics, mathematical ecology, or most of economics. In the case of economics, attempts at macroeconomic forecasting do not show much evidence of empirical adequacy greater than might be expected from astrology or the conscientious examination of entrails. Perhaps population genetics can claim some empirical support for the Hardy–Weinberg law, but this will not take us very far toward an empirically adequate account of the dynamics of real populations. By contrast, history, aerofoil design, and violinmaking are highly empirical activities.

A particularly interesting case in this context is Hull's chosen example of systematics. On the one hand phenetics and the offshoot of cladistics, pattern cladism, which, as Hull so nicely documents, comes close to methodological reconvergence on phenetics, are evidently motivated in large part precisely by the goal of firm grounding in empirical fact. What is striking is the apparent connection of this goal with the explicit rejection of theoretical assumptions. One can almost read into Hull's history of systematics the conclusion that one can have empirical grounding or theory, but not both. Even if this is too extreme a conclusion, the following question is pressing: Do the debates about the appropriate theory and methodology in systematics that Hull describes depend in any serious way on evaluation of evidence? As far as I can see the answer is firmly negative.

Consideration of the scientific status of systematics leads me to the second of my suggested criteria, that science should be cooperative and cumulative. I conjoin these properties because it is clearly part of Hull's thesis that the way scientists are able to build on the work of their colleagues and predecessors is part of the explanation of science's success, of its capacity to provide successively more-adequate solutions to the problems within its domain. There is no doubt that in a very broad sense, which includes backstabbing and attempted professional assassination, Hull shows systematics to be cooperative. It includes, that is to say, a great deal of professional interaction, some of which is cooperative. Whether it is cumulative, on the other hand, is much less clear. Of course we need not doubt that in the straightforwardly empirical sense of listing more species, finding more reliable diagnostic criteria, and so on, systematics has made steady progress. But this is not the kind of change that Hull describes. On the contrary, the more-theoretical debates that he does describe seem to circle around inconclusively in the manner often attributed to traditional philosophical problems rather than advance in any intelligible direction. Certainly there is no indication of convergence on any broad theoretical consensus. Although perhaps the cladists are seen as currently occupying a somewhat dominant position, Hull's narrative leaves entirely open the possibility that within a few years the pattern cladist wing will have reinstated essentially the position being staked out by pheneticists at the beginning of the story. Moreover, not only do the cooperative parts of systematics not seem especially cumulative, but the cumulative parts do not seem very cooperative. Few things, I suspect, provide a more isolated scientific avocation than developing an improved classification for, say, the Compositae of Antarctica. Ernst Mayr, in his introduction to *Birds of the Southwest Pacific* (1945), remarks that no book on this subject exists in the English language. Although the work includes a scattering of references to technical papers, there is no bibliography or author index. I suspect a comparison of the frequency with which this work is cited to the citations of, say, Mayr's rather more theoretical *Animal Species and Evolution* (1963) would reinforce the distinction I am suggesting. At most a tiny handful of colleagues will have any competence to criticize a work such as the former. I do not mean to suggest that such labors start entirely from scratch; but then no interesting human activities whatever do that.

Turning to my third suggested criterion, the important role of broad theoretical principles, my point is again that the role and status of these will vary greatly from one area of scientific inquiry to the next. In large parts of the physical sciences such theories often seem to have a dominant role. Mathematically precise systems of equations appear to define whole fields such as quantum mechanics, electrodynamics, or physical chemistry. A casual glance might suggest a similar structure to fields such as microeconomics or population genetics. But I suggest that such an assimilation would be misleading. The problem with these latter fields is that there is little reason to suppose that there are any real systems that even approximately realize the mathematical formalisms of such disciplines.[10] Further along the spectrum of scientific disciplines, the role of grand theories becomes even more attenuated. Perhaps overarching theories of sociology, human psychology, or meteorology will eventually be accepted. But I see no a priori reason to anticipate or even welcome such developments.

The main point of the preceding discussion is to cast doubt on what I have taken to be a presupposition of an ambitious interpretation of Hull's thesis, that there is some reasonably homogeneous class of human practices that merits the honorific title "science." I have argued that the relevance of all the major features that might plausibly be taken to differentiate science from more or less related human activities varies greatly from one scientific activity to the next. Perhaps an activity in which all the features I have discussed were strongly evident would necessarily be a science. But surely to require this would be to exclude the majority of activities that currently lay claim to that title. Science, to borrow an important idea from the later Wittgenstein, is best seen as a family resemblance concept. That is, there will be a number, perhaps an indefinite number, of features characteristic of parts of science, and every part of science will have some of these features, but very probably none will have all. Consequently, it seems likely that the relevance of *any* mechanism of the kind that Hull describes should be expected to vary greatly from one area of science to another, and the observation that sciences are social processes will not provide a possible source of scientific unity. Science is not *a* process.

Toward a Pluralistic Epistemology

Science, construed simply as the set of knowledge-claiming practices that are accorded that title, is a mixed bag. The role of theory, evidence, and institutional norms will vary greatly from one area of science to the next. My suggestion that science should be seen as a family resemblance concept seems to imply not merely that no strong version of scientific unity of the kind advocated by classical reductionists can be sustained, but that there can be no possible answer to the demarcation problem. But as I have also indicated, I find Popper's motivation for attending to this question to be entirely convincing.[11] Although I do not share Popper's particular hostilities to Marxism and psychoanalysis, which provide the negative test cases for his account, I do think it is vitally important, especially given the social and political power that presently accrues to successful claimants to the title of scientist, that we develop critical principles for assessing the validity of such claims. It would strike me, for example, as a fatal flaw in my position if it led to the conclusion that nothing could be said in explanation of the epistemic superiority of the theory of evolution over the apparently competing claims of creationists.

While it is clear that my Wittgensteinian thesis does indeed exclude a simple criterion of demarcation, I offer instead, and once more, the consolations of pluralism. I suggest that we try to replace the kind of epistemology that unites pure descriptivism and scientistic apologetics with something more like a virtue epistemology. There are many possible and actual such virtues: sensitivity to empirical fact, plausible background assumptions, coherence with other things we know, exposure to criticism from the widest variety of sources,[12] and no doubt others. Some of the things we call "science" have many such virtues, others very few. To take the example mentioned above, it will hardly be difficult to demonstrate in such terms the greater credibility earned by the subtle argumentation and herculean marshaling of empirical facts of a Darwin, followed by a century and more of further empirical research and theoretical criticism, than that due to the attempt to ground historical matters of fact on the oracular interpretations of an ancient book of often unknown or dubious provenance. Such an approach would at the very least have the capacity to capture the rich variety of projects of inquiry, without conceding that anything goes. And to foreshadow the topic of the next chapter, it could provide an epistemological standard for science that would be overtly and unashamedly normative.

One further consequence of my position should be mentioned. Earlier in this chapter I distinguished two possible conceptions of the demarcation problem: the differentiation of good from bad science, and the distinction of science from nonscience. The schematic suggestions in the previous paragraph gesture toward a solution to the former problem. But the approach I advocate also implies that a solution to the second problem is unlikely to be forthcoming. Many plausible epistemic virtues will be exemplified as much by practices not traditionally included within science as by paradigmatic scientific disciplines. Many works of philosophy or literary criticism, even, will be more closely connected to empirical fact, coherent with other things we know, and exposed to criticism from different sources than large parts of, say, macroeconomics or theoretical ecology. In general, I can imagine no reason why a ranking of projects of inquiry in terms of a plausible set of epistemic virtues (let alone epistemic and social virtues) would end up with most of the traditional sciences gathered at the top. No sharp distinction between science and lesser forms of knowledge production can survive this reconception of epistemic merit. It might fairly be said, if paradoxically, that with the disunity of science comes a kind of unity of knowledge.

11. Science and Values

Science is a human practice. It is a practice that consumes the lives of many, and affects the lives of all. If the issues that have been addressed in the preceding chapters have real importance, it is because they are relevant in some way to how this activity should be encouraged, or to how much influence it should have on the way human life is carried on. A major aim of this final chapter is to claim that these issues do, indeed, have this kind of significance. Thus I arrive at the most important topic in the philosophy of science, the relation of science to human values. What contributions can or should science make to human well-being? Under what circumstances is it most likely to make such contributions? To what extent do the more local or personal values of scientific practitioners affect the products of science? The difficulty of such questions can be seen in the fact that their resolution would require a full understanding of both the nature and sources of human value, and the determinants of scientific belief. Thus my remarks on this topic must be partial and provisional. I shall, however, in addition to a brief account of the significance of some specific doctrines that have been criticized in the preceding pages, defend the importance of a more critical approach to scientific doctrine than has been common in most recent philosophy of science. The best way of establishing this point would be to provide a detailed critique of some significant part of science. Such a task must await another volume. But I do believe that the conclusions reached in these pages go some way toward establishing the importance of such critical essays.

If this discussion is not to be question begging, a broad conception of human values will be necessary. In particular, I am certainly willing to allow that the accumulation of knowledge for its own sake might be a central human value. One conception of the relation of science to values is that the sole value to which science should aim to contribute is the acquisition of knowledge, or at least true belief. It must be admitted that this will allow the possibility of secondary effects, as scientific knowledge facilitates the production of vaccines and nuclear weapons, refrigerators and toxic waste dumps. But these secondary effects, on this view, are not the responsibility of science. The more we know, the more we can do both good or bad. What choices we make about our use of knowledge are no more the responsibility of the providers of knowledge than it is the responsibility of farmers if we decide to overeat.

Although the foregoing remains a common, perhaps even a dominant, view of the normative status of science, it is a view that has become increasingly untenable: it has become widely agreed that values enter into the process of science long before decisions are made about technical applications of scientific knowledge. To begin with, however objective the methods by which a problem is investigated may be, no one has proposed a method internal to the processes of science by which the choice is made which problems are even exposed to these processes of investigation. My conclusions in earlier chapters are relevant to this point. For if, as I have argued, there are many threads of partial order in the universe, but none with any overriding or privileged significance, the choice of fields of investigation may be wide indeed. And some problem areas that are chosen for intensive exploration might lack any genuine preexisting order to be disclosed.

But while the preceding point shows how external values might help to determine the output of science, it does not suffice to establish the possibility that the *content* of scientific claims might be affected by social or personal values unrelated to the processes of science. The possibility of such influences on the content of scientific belief does, however, start to become apparent in the light of a point introduced in the preceding chapter, that all scientific reasoning involves a range of assumptions that are not themselves subject to empirical confirmation or rejection in the process of scientific inquiry. Exactly what kinds of values do come to bear on such background assumptions, and the circumstances under which they do so, are questions that can be addressed adequately only by detailed work in the history of science. Presently I shall consider the way in which the philosophical positions addressed in earlier chapters provide underlying assumptions with considerable normative significance to certain areas of scientific inquiry. Before doing that I shall attempt to give some sense of the wide range of factors that have been found relevant to understanding scientific belief. To that end I shall consider briefly an episode in the history of science that has recently been brilliantly redescribed.

The episode I have in mind is the debate between Hobbes and Boyle as described in Steven Shapin and Simon Schaffer's *Leviathan and the Air-Pump* (1985). This debate, while focusing directly on Boyle's experimental use of a piece of machinery that he claimed was capable of producing a vacuum or at least a partial vacuum, also involved explicitly contrasting conceptions of natural philosophy. The episode is of particular interest because the Whiggish tradition in the history of science has tended to treat Boyle, whose views seem quite congenial to contemporary scientific ideology, as so clearly correct that Hobbes's views are not worthy of serious consideration. (Shapin and Schaffer clearly document the extent to which scholars have ignored this aspect of Hobbes's work.) But a more open-minded analysis of the issues involved in the debate suggests that there was considerable weight on both sides, and that Boyle's success in establishing his conception of knowledge production owed more to political negotiation in the complex sociopolitical environment of Britain during the Restoration than to the inherent superiority of Boyle's philosophical or empirical position.

To lead to the most significant consequences, the thesis that scientific belief depends always on background assumptions not all of which can be equally subject to empirical warrant must apply not merely to more or less theoretical beliefs, but also to what are taken to be the empirical data on which scientific beliefs might be founded. If it does not, it provides no more than a powerful motivation to erect a sparser theoretical structure on truly empirical foundations. But if it does apply to the data themselves, this is tantamount to the denial that there could be any *foundations;* for even the putative foundations are seen to rest in part on some higher aspect of the edifice. Although there is a good deal of talk nowadays about the "theory-ladenness of observation," it is my impression that the idea behind such talk is not always taken quite seriously.[1] One of the most significant points that Shapin and Schaffer demonstrate with their account of the epistemological debates in seventeenth-century Britain is that there is a much broader sense in which data are theory-laden than those that are more commonly discussed: namely, that it is a matter entirely open to debate what kinds of observations should be taken to provide the canonical data for natural philosophy, or science.

Boyle and the other members of the incipient Royal Society were concerned to promote a particular paradigm of the canonical datum, of what could serve as foundational for scientific knowledge. The data in question were to be produced with the help of scientific instruments held to be capable of penetrating more deeply into nature's secrets than could unaided observation. One obvious exemplar of this possibility was the telescope; that being specifically promoted by Boyle was the air-pump. The use of such instruments would also require the existence of a class of specially trained and expert observers capable of operating the instruments correctly. We can see here essential ingredients of the professionalization of science. More generally, we can see the antecedents of the conception of science, and thus of authoritative knowledge production, that dominates our own society: highly trained professionals, often equipped with extremely elaborate and expensive instruments and laboratories. This model of knowledge production is so familiar that it is liable to seem inevitable. By revealing a real debate about its legitimacy before its achievement of the dominant position it now has, we can begin to see that, whereas it has undoubtedly produced considerable epistemological achievements, it is not unique, and it also has the capacity to serve certain social interests rather than others.

What advantages can be claimed for the laboratory equipped with complex instruments as a privileged source of knowledge of nature? Nowadays we are very likely to answer in terms of the technological achievements that have come from laboratory science. But to do this would only confirm one of Hobbes's major accusations against the experimentalists, that they were degrading philosophers to the level of "apothecaries and gardeners . . . [and] many other sorts of workmen,"[2] as deplorable intellectually as socially. The force of this complaint is amplified by the fact that Boyle himself distinguished sharply between the matters of fact that, he believed, could be demonstratively produced by his machinery, and the causal knowledge that he generally eschewed. But the latter, according to Hobbes, was the only knowledge worthy of the attention of a true philosopher. As I shall argue in a moment, Hobbes is surely right in rejecting this aseptic conception of empirical knowledge.

Objectivity, an obvious desideratum of "matters of fact," has generally and appropriately been associated with some kind of intersubjectivity or consensus. Thus Boyle attaches great weight to the replicability of the experiments that were carried out in his laboratories. Moreover, experiments could be performed in front of numbers of witnesses and, more important still, qualified and respectable, and therefore reliable, witnesses. But these claims are not beyond dispute. Replicability, both for Boyle and for contemporary experimentalists, can seldom be guaranteed, and in fact proved very difficult to achieve with Boyle's temperamental and unreliable air-pumps. Indeed it is plausible that reliable replicability is something that happens only when science becomes technology. And the publicity and openness of this method of generating agreed-on matters of fact were obvious targets for Hobbes. Whether or not laboratories are technically public spaces—which in the present age of military funding of science they increasingly are not—they are not spaces that are in fact either accessible or intelligible to the public. And to be an insider in the laboratory, an authoritative witness or an operator of the equipment, is a privilege accorded only after lengthy training and indoctrination in the official ideology of the laboratory. Hobbes insinuates further that the laboratory is a very hierarchical place, in which a small few are much more authoritative than the rest—another suggestion that is surely prescient of the subsequent development of laboratory science.[3] Thus Hobbes is surely on firm ground with regard to this aspect of objectivity when he claims that "mean and common experiments" are "a great deal better witnesses of nature than those that are forced by fire, and known only to a very few."[4] Boyle's matters of fact are seldom fully replicable, and almost never publicly accessible.[5] In short, the general conception of privileged "matters of fact" developed throughout the course of modern science is one open to serious challenge.

At a more fine-grained scale, the debate between Boyle and Hobbes over the interpretation of the experiments with the air-pump excellently illustrate the role of background assumptions in the formulation of data. This is the more familiar sense of the "theory-ladenness of observation." The interpretation of all Boyle's experiments with the air-pump depended crucially on whether one thought that the pump had some tendency to evacuate the container to which it was connected. For Hobbes, committed to a version of plenism, the production of a vacuum was not in question; and it was not hard to provide hypotheses of how and where the air was finding its way back into the container. Hobbes also had a conception of the nature of air that sufficed to explain many of Boyle's experimental results. Naturally occurring air, he held, was a mixture of pure air ("aer purus") and earthy matter. The operation of the pump would enrich the proportion of pure air, since only this, being infinitely divisible, would find its way back into the apparatus. The peculiar behavior and motions of this pure air could then be deployed to explain the phenomena produced in Boyle's alleged vacuum.

It would be possible to object that the *data* are not the phenomena, such as the equalization of Torricellian barometers or the bursting of bladders *in a vacuum,* but merely the occurrence of such phenomena in an apparatus of such-and-such a kind. Such a maneuver is greatly facilitated by Boyle's distinction between matters of fact and causes. But Hobbes was surely right to reject this distinction. That a certain phenomenon can be produced by a certain experimental setup is the kind of claim that deserves precisely Hobbes's devaluation as mere artifice. It is the claim that, for example, the arms of a Torricellian barometer are of equal heights *in a vacuum* that has scientific consequence. And such a causally laden claim can never be a direct observation, least of all in a complex experimental apparatus.

According to Shapin and Schaffer, however, such ontological issues as the nature of air and the possibility of a vacuum were of only marginal importance in the debate between Hobbes and Boyle. The real divide between them concerned the appropriate "form of life" of the natural philosopher, a divide rooted in the political instabilities of seventeenth-century England. Hobbes, extending the anticlericalism that is a well-known feature of his political philosophy, considered that any closed group of intellectuals with substantial independent authority would tend to be driven by their own political interests rather than by the disinterested pursuit of knowledge claimed by the experimentalists. This opinion, encompassing the view that knowledge is constructed by social agents with political interests that that knowledge is apt to serve, reveals Hobbes as a precursor of much late twentieth-century thinking about science. This much, at least, must surely be conceded to Hobbes's prescience: the experimentalist conception of science the beginnings of which he criticized has subsequently grown into a social institution with every bit as much power as the ecclesiastical establishment that Hobbes perceived as such a threat to social and political well-being in the seventeenth century. This development does not exclude the possibility that this power is benign, although perhaps it makes it unlikely that it is entirely so.

An optimist about the benignity of science might maintain that despite the great power that has undoubtedly accrued to the institutions of science, their singular virtue is that they are structured in such a way that this power will always be channeled in socially, or at least epistemically, useful ways. This is the basic strategy of Hull's account of science as a process set up in such a way as to generate progress despite the worst intentions of its practitioners. One of my central criticisms of Hull's account was that it did not leave space for evaluation in any normative terms of the lineages of scientific ideas he postulates. I shall now offer one example of such a normatively, though not necessarily epistemologically, objectionable lineage. This should suffice to illustrate the possibility

that projects of inquiry are vulnerable to straightforwardly political criticism.

A main tradition of development economics has been concerned to discover the determinants of household income, with
one important goal being the discovery of those economic changes
that will effect increases in such incomes that are appallingly low.
The concept "household" is difficult to define in a satisfactorily
cross-cultural way, and no doubt interesting lineages could be
described for the evolution of this concept. Recently a number of
feminist critics have pointed out that this is often an ill-chosen
goal: in many cases increasing the income of a household will
have no effect, or even a negative effect, on the incomes of its
poorest members, women and children (Dwyer and Bruce, 1988;
Bruce, 1989). Moreover, the whole problem is greatly exacerbated
by the difficulties in defining what is meant by income. Ideally,
one would like to include everything that contributes to the well-
being of the people constituting the household, but in the real
world (or so it is argued) practical difficulties preclude doing so.
What tends to happen is that income is equated with paid labor,
and a great deal of productive work that occurs within the household is ignored. Unfortunately this divide tends to correspond
closely to that between work done by men and work done by
women, allowing massive discounting of the contributions by
women to social welfare (Waring, 1988, especially chap. 3). Such
a concept of income almost inevitably leads to policies directed at
income enhancement that are not perceived by women as increasing their well-being.

It will be no surprise that such a normatively crucial question
as the estimation of human well-being should give rise to problems
of this sort. Yet none of this reveals any factual error in the history
of investigation of the determinants of household income, provided that it is understood that quite specific definitions of these
concepts are inevitably involved. And consequently not even the
most optimistic account of the tendency of scientific investigation
to progress toward objective truth will show how any error of
the kind indicated by the normative criticisms outlined above must
be disclosed. The reason such a case provides a difficulty for Hull's

project is that whereas "household" as it applies to an actual historical economic concept may well be treated as a conceptual lineage, "household" as it appears in contrast with individual, gendered, members of households must be kept available as an ahistorical general term of critical analysis. The overriding reason for this insistence is that debates concerning the normative dimension based, for example, on different views on the appropriate roles of men and women in households will drive such a project of inquiry in ways that are largely independent of the empirical facts concerned.

It will perhaps be unsurprising to all but the most fanatical defenders of scientific objectivity that investigations so directly concerned with normative matters should show this sensitivity to normative disagreement. (This is no doubt why an unbridgeable gulf between positive and normative economics is generally emphasized at an early stage in the pedagogy of economics, to insist that the scientific status of the former is not in danger of contamination by the latter.) However, with the disclosure of the existence of a route by which such extraneous matters can influence the content of scientific belief, it will be difficult to exclude such intrusions into less likely areas of science. I shall next indicate how the metaphysical doctrines that have provided the central focus of this book can affect scientific belief—in a number of areas—in ways that have considerable normative, or political, consequences.

Essentialism and Politics

Essentialism is one of the clearest examples of a doctrine that can provide a necessary presupposition of various scientific enterprises, and a presupposition that in many cases is highly politically contentious. Essentialism, as discussed at length in Part I, relates to the significance of the classification of things into kinds. The thoroughly antiessentialist position I advocated there mandates serious skepticism about the extent to which homogeneity among the members of any kind should be assumed. In many cases concerning human beings, the very process of classification carries dangers. One example of such possible dangers, much discussed in light of Foucault's *History of Sexuality* (1979), is the case of categorization of people according to their sexual practices. Thus Foucault, as is well known, argued that the categories of homosexual and heterosexual were invented some time around the second half of the nineteenth century. What had previously been regarded as kinds of behavior, exhibited at various times by various people, came rather to be seen as defining kinds of people.[6] During the last century this has come to seem to many an entirely natural way of thinking about human identity. And it has many consequences. Of most obvious significance to my concerns here, it licenses the search for further characteristics of the supposed kinds, and for the underlying (perhaps even essential) features (genes, hormones, and so on) responsible for those characteristics. Indirectly this may help to legitimate behavioral stereotypes and to reinforce the assumptions of heterosexuality as the "normal" type and of homosexuality as pathological.[7] But given the arbitrariness of the decision to treat homosexuals and heterosexuals as *kinds* of people in the first place, this whole enterprise may be no more firmly grounded than the search for the typical characteristics and genetic peculiarities of stamp collectors or aficionados of crossword puzzles. In cases with this level of social significance, the categorial empiricism I advocated in Chapter 3 becomes a political imperative.

The problems to which essentialism gives rise have been treated in the greatest depth by feminist scholars in various disciplines concerned with essentialist conceptions of men and women.[8] I have already discussed and criticized such conceptions at some length in Chapter 3. Here I need only emphasize that essentialist assumptions in the analysis of gender difference are not merely philosophically and scientifically suspect but also have enormous potential for political abuse. The general way in which this can happen is very simple. When differences in the behavior of men and women are explored against the background of the model of distinct biological kinds, it is inevitable that any systematic differences that are found will be taken to be explained, or explicable, in terms of the intrinsic differences between members of the two kinds. Thus gender difference will be shown to be biological, and hence natural; and even if this does not quite entail that they are immutable, at least it will raise doubts about the wisdom of attempting such changes. So, for instance, E. O. Wilson (1978) asserts that "there is a cost, which no one can yet measure, awaiting the society that moves either from juridical equality of opportunity between the sexes to a statistical equality of their performance in the professions, or back toward deliberate sexual discrimination" (p. 153). This is because "even with identical education for men and women and equal access to all professions, men are likely to maintain disproportionate representation in political life, business, and science" (p. 138). Evidently these claims amount to a serious attack on the political aspirations of the contemporary feminist movement. What makes them particularly insidious is that they are liable to sound not so much like a political move but like cold, hard scientific facts announced regretfully by a distinguished Harvard biologist. But when it is seen how scientifically worthless such claims are, they can be exposed for what, intentionally or otherwise, they are, which is antifeminist politics mobilizing the epistemic authority of science in their support. A central step in exposing the lack of any scientific merit to such claims is to undermine the essentialist perspective that makes it plausible to interpret statistical differences between the behavior of men and women as simple consequences of intrinsic differences between members of those kinds.

Reductionism and Politics

Essentialism, I have suggested, helps to legitimate scientific projects of inquiry and conclusions that are both fraught with potential political abuses and devoid of legitimate epistemic credentials. Reductionism, on the other hand, tends rather to exclude or to delegitimize lines of investigation that do not meet the exacting conditions of scientific knowledge presupposed by the reductionist program. Clearly, if I am right in my rejection of the legitimacy of this program, this arbitrary restriction on the concession of epistemic authority may also be open to political abuse.[9] Thus, for example, reductionist arguments are readily deployed skeptically against explanations of social problems in sociological terms (unemployment; inadequate, overcrowded housing; and so on) in favor of some hypothetical but as yet nonexistent explanation in terms of features of individuals (genetic deficiencies, mental illness, and so on), and thereby can be deployed against programs intended to alleviate such problems. However, although such uses may occur, I think that the political significance of reductionism is generally more indirect.

First, reductionism is a crucial component of the defense of scientism, thereby functioning at a level similar to the issues debated between Boyle and Hobbes about the general authority of laboratory science rather than mandating any particular use of that authority. Many humanistic, scientific, and other approaches to the understanding of human or social problems have been advocated by various people. But reductionistic understanding (in the sense considered in this book) is a prerogative of science. If the reductionistic account of science mirrors the hierarchical structure of nature, then science will be the one source of truth in these as in other matters.

Second, reductionism interacts with essentialism by justifying a particular approach to the elaboration of the human kinds considered in the previous section. Thus the assumption that men and women form natural kinds might be relatively innocuous if the kinds were taken to be *sociological*. One might suppose that social forces were sufficiently homogeneous to provide rather homogeneous behavior among many human groups. But this would not imply anything about the malleability of this behavior in response to political action aimed at diverting these forces.[10] It is quite another matter if these are understood as natural kinds in the essentialist sense addressed in the preceding section, as is typical in the context of a reductionist science. The behavioral correlations characteristic of human kinds are to be explained in terms of genes or hormones, and if the behavior is mutable at all, it is so only at the cost, emphasized by Wilson, to be incurred for forcing human beings from their "natural" conditions.

Finally, reductionism, as I have emphasized in earlier chapters, is intimately connected with determinism or broader assumptions of causal completeness. But this deserves a separate section.

Determinism and Politics

Both essentialism and reductionism, I have argued, tend to explain the actual states of things in ways that present those states as natural, and changes from those states as, at best, difficult to achieve and exacting some cost. Thus, in application to human affairs, they are well suited to the legitimation of conservative politics and to the discouragement of proposals for significant social change. The same is true, for rather different but more familiar reasons, of determinism. Global determinism, at least, implies that the present state of things is not only natural, but inevitable. I do not propose to enter into the ageless debate as to how, if at all, commitment to the inevitability of everything that happens is consistent with a belief in the possibility of the amelioration of social—or even individual—conditions through deliberate effort. But if my conclusions at the end of Chapter 9 are correct, the kind of disorderly metaphysics I have been advocating is surely better suited to progressivism than to any position involving determinism. In this context it makes little difference to replace determinism with an indeterministic version of causal completeness; this replacement leaves it equally problematic how intentional progressivism can increase the probability of progress. On the view I have defended, on the other hand, it is quite possible on the social, as on the individual, level that ameliorative efforts may be capable of creating forms of social order where none existed before.

This issue about determinism has relevance even to what I have so far allowed as harmless, determinism as a methodological principle. I do not, of course, wish to criticize the principle that it is a worthwhile endeavor to attempt to explain why apparently similar events produce different sequelae, although it has been a major claim of this book that such an attempt may be doomed to failure. However, it is a small step from this inoffensive maxim to the position that the promise to provide as much explanation of different outcomes as possible gives a reason to choose one particular approach to a certain domain over less deterministic competitors. Such a principle would be justified only by the commitment to some kind of actual determinism. The bearing of such partly methodological and partly metaphysical issues on actual scientific research is brought out clearly by Helen Longino (1990, especially chap. 7).

Longino compares two very different approaches to the explanation of human behavior, specifically to the explanation of gendered differences in behavior. The first is linear and seeks correlations between perinatal hormonal conditions and behavior later in life. The assumptions underlying this program are that concentrations of hormones affect the development of the brain, and that states of the brain affect later behavior. Although no one, presumably, thinks that every detail of subsequent behavior is explicable in this way, the hope is that where there are systematic and substantial differences in male and female behavior, these will prove traceable to the relevant perinatal conditions. In the present context a point also central to Longino's discussion is crucial: if the program were to be fully successful it would show no role for the choices of anybody, male or female, in the display of any of the behavior explained by such research. In contrast, Longino describes a program initiated by Gerald Edelman and Vernon Mountcastle, the "group selective theory," the details of which need not be discussed here. A crucial feature differentiating this approach from the linear model, in which behaviors are caused once and for all by events early in development, is that it is dynamic: the states of the brain being described are subject to continuing alteration in interaction with the environment. Moreover, it does not aim to explain specific kinds of behavior, but rather addresses the question, as Longino puts it, "What sort of structure and functioning must characterize a brain capable of long- and short-term memory, learning and correction of memory, observational as distinct from conditioned learning, self-awareness, creativity, and mediation of action and experience?" (1990, p. 147). This contrast dovetails strikingly with a general thesis from my discussion of reductionism: we should expect structural analysis to tell us about the capacities of an entity rather than about its actual behavior. The group selective theory clearly meets this expectation, whereas the linear model equally clearly fails to.

A number of further points are brought out by this contrast. First, the explanatory goals of these programs are so different that it is quite unclear how we could compare their scientific credentials. Perhaps a few decades of hindsight may reveal that one has been a great deal more successful than the other, but even that might show only that the less successful was more difficult to implement. On the other hand the different background commitments and assumptions underlying the two programs might be quite sufficient to motivate a preference for one over the other (whether by an individual researcher or by a funding agency). Longino makes it quite clear that her sympathies lie with the group selection model. As a feminist, she is committed to the position that the personal and political decisions of individuals have some chance of changing the kind of behavior, especially gender-specific behavior, that is exhibited. And whereas this kind of possibility is part of what the group selection model is intended to explain, it is unintelligible to, or at least beyond the scope of, the linear model. It is surely entirely reasonable for anyone to pursue or promote the kinds of research programs that have some chance of explaining the kinds of things they take to be true. And the theoretical choice presently under consideration is one that appears to depend on considerations at that level of generality.

The example of such linear—I am tempted to say simplistic—explanations of gender-specific behavior is admirably suited to reveal the relevance to scientific theory choice of all three of the philosophical positions I have been considering, and it is thus no accident that feminist critiques have been so important in undermining the scientistic metaphysics that has been the topic of this book. The very formulation of the questions, about the cause of different styles of play between boys and girls, sexual aggressiveness, career choice, and the like between men and women, is grounded in a basically essentialist conception of men and women. The answers that the linear model proposes are paradigmatically reductive: what appears to be a set of issues at the level of human behavior is to be understood solely in terms of events at the chemical and physiological levels. And finally, the broader assumptions about the explanation of human behavior of which it is naturally a part are deterministic. If the suggestion I made at the end of Chapter 9 about free human action is accepted, then it will be clear that something more like the group selective model should be preferred.

The goals of this and the preceding two sections have been quite limited. I have not tried to provide a case that theory choice is underdetermined by strictly internal considerations, nor that the essentialist, reductionist, and determinist assumptions I have been considering are false. The latter has been the goal of the preceding eight chapters; the former is widely agreed among students of the sciences, and its denial would be extremely difficult for anyone persuaded by the negative metaphysical theses of earlier chapters. If, moreover, human action is causally efficacious, so that the structure of things is partly determined by our own choices, then there will be different patterns of things to explore corresponding to the different projects we have for making the world the ways we want it to be. What I have tried to show is that these metaphysical theses have considerable significance for choosing between different ways of producing knowledge. What is more, I am convinced that rejection of these theses would make us more likely to pursue projects more conducive to human welfare.

Politics and Disorder

The debate between Hobbes and Boyle, according to Shapin and Schaffer, was above all concerned with social order. Whether or not this claim is fully accepted by future historians, there can be no doubt that order of various kinds is central to the function as well as to the assumptions of science. It used to be very common to assert that the hallmarks of science were prediction and control. And although prediction has run into some philosophical hard times lately, control, or the imposition of order, remains a central function of science. And by claiming this as a central function, I mean to imply that it is a function that serves, in part, to justify the expenditure of vast social resources on science.

Science aims both to detect order and to create order. The great achievement of early astronomy was to disclose the order concealed by the apparent complexity of celestial events. Clearly, order in this domain was not going to be created by human efforts. Chemistry and terrestrial physics look like more promising candidates for projects of creating order, although it is debatable whether science contributed more to technology or vice versa for much of the early modern history of the physical sciences. However, what I want to focus on here is recent science, and generally nonphysical science. Here the relations between the discovery and the construction of order are complex and easily misrepresented.[11]

Where there are background assumptions that affect, but are not determined by, the empirical content of a discipline, we should see that discipline as engaged, at least in part, in constructing a kind of order that is defined by that set of assumptions. Economics again provides a good illustration. Many of the key terms of economics are, as I argued above for "income" and "household," open to debate as to how they should be defined. And the kind of debate in question can involve substantial normative disagreement. Consider one further example, "work." Work is some subset of human activity that excludes leisure, and distinctly, pure idleness. Exactly which activities count as work is quite clearly not an empirical question. Are productive hobbies, such as gardening, carpentry, and so on, work? Are drug-dealing and gun-running kinds of work? Are paid activities that are quite unproductive, such as running an advertising company, or perpetrating arbitrage, work? These are matters to be decided, not discovered. But when economists claim lawlike relations between unemployment and inflation, for example, unemployment must be understood in relation to work, which must be defined in one way rather than in many other possible ways. If economists are successful in controlling—to some degree—unemployment, what they are controlling will depend on such definitional decisions. And these, in turn, will depend on what is admitted to be valuable or important.

I cannot attempt to develop these views about economics in much detail here. I am, however, currently undertaking a detailed study of economics in the light of the views developed in this book, and the extent to which economics serves to create rather than to disclose economic reality will be a major focus of that project. For now, the crucial point is that the view just sketched is one that becomes very natural given the disordered universe, or perhaps more accurately, the universe with diverse strands of partial order, that I have tried to describe; whereas it makes little sense given the assumption of unitary cosmic order that I have been criticizing. In particular, the pluralism of kinds defended in Part I should make the indeterminacy of the categories with which a domain can be approached unsurprising. Perhaps it is relatively easy to believe of economics that there is some degree of arbitrariness in the classification of phenomena. I have argued at length that the same is true of biological classifications. I find it very plausible that the same can be said even of the physical sciences, although that is not something I have attempted to defend. Any of these actual or possible systems of classification might form the basis for a project of inquiry or a program for controlling some aspect of nature or of society. Which among such projects we choose to pursue may depend on many factors, but principal among these will be political factors.

Such scientific pluralism will be much less plausible if one thinks that the kinds of things there are are antecedently and uniquely given by nature, and thus the legitimate projects at least of *scientific* inquiry will be very few. It will, finally, be clear by now how the commitment to the disunity of science is integrally connected to this position. For if science were unified then the legitimate projects of inquiry would be those, and only those, that formed part of that unified whole. On the picture I am presenting, only a society with absolutely homogeneous, or at least hegemonic, political commitments and shared assumptions could expect a unified science. Unified science, we might conclude, would require Utopia or totalitarianism.

Relativism, Realism, Pluralism

Let me conclude by relating the view of science that I have tried to describe in the preceding pages to some more familiar contemporary ideas. First, is my position a version of relativism? Probably in common with the majority of accounts of science, the answer to this must be yes and no. I have argued (mainly in Part I) that the theories that science comes up with must depend on (that is: will be relative to) the purposes for which they are intended, and moreover (mainly in this and the preceding chapters) that they will depend on many other assumptions, most importantly, political ones. If this is relativism, then I am happy to admit to it. Often, however, relativism is taken to involve much more, for example, the denial that any scientific account could be true, or the insistence that all scientific theories have equal epistemic credentials and equal value. One particularly influential statement is the "equivalence postulate" of Barry Barnes and David Bloor, that "all beliefs are on a par with respect to the causes of their credibility" (Barnes and Bloor, 1982, p. 23). As a methodological maxim for the study of particular scientific practices or the history of science this seems quite unexceptionable; the movement against seeing the history of science through the lenses of our current scientific beliefs is entirely salutary. Nevertheless, such statements have been widely understood as implying that all scientific beliefs are of equal worth. With this, as was made clear in the previous chapter, I wholly disagree. My view, on the contrary, is that scientific practices vary from the highly plausible to the quite incredible and, often in a parallel fashion, from the extremely valuable to the entirely pernicious. My concluding remarks in the last chapter give a sketch of my view of the epistemic spectrum; in this chapter, I have tried to give some sense of the normative.

What, then, of realism? I referred to the pluralism defended in Part I as "promiscuous realism," and the term *realism* is intended seriously. Certainly I can see no possible reason why commitment to many overlapping kinds of things should threaten the reality of any of them. A certain entity might be a real whale, a real mammal, a real top predator in the food chain, and even a real fish. Many, perhaps all, of these designations might be the appropriate characterization of that object for some legitimately scientific purpose. I do not see why realism should have any tendency to cramp one's ontological style. Reflecting the continued dominance of microphysics in the philosophy of science, realism is often taken to concern solely whether we should credit the particles described by microphysicists. Since microphysics seems to me a rather distant suburb on the intellectual—or even the ontological—map, I certainly do not take any significant part of my philosophical position to rest on the answer to this question. While remaining largely agnostic, I do find myself impressed by Ian Hacking's remark about electrons: "if you can spray them then they are real" (1983, p. 23).

Finally let me mention the philosopher with whose general

perspective on science I find myself most closely in agreement, Paul Feyerabend. Feyerabend's assessment of the social and political position of science in contemporary society (1978, part 2) strikes me as an oasis of serious critical analysis on a topic that, astonishingly enough, has been almost entirely ignored by philosophers. Feyerabend's notorious epistemological anarchism is intended above all as therapy against the antidemocratic and oppressive consequences of the monopoly of epistemic authority maintained by science (1975, p. 17). Although Feyerabend's acute sense of the sociological unity of science sometimes makes him write as if science were a unified project, it is clear that his general epistemological pluralism and his attack on philosophers' accounts of scientific methodology constitute, among other things, devastating attacks on various attempts to construe science as unified.

There are, however, important matters of strategy on which I diverge from Feyerabend. Feyerabend emphasizes the impossibility of explaining scientific advances in terms of canons of rationality or scientific method, and the way in which epistemic practices outside the mainstream of science are ignored and dismissed rather than subjected to any serious criticism. The conclusion is that there is nothing unique about scientific practices beyond their authority and, perhaps, arrogance. I want to reach the same conclusion by means of a slightly different premise: there can be nothing unique about science, because there is nothing common to the various domains of science.

But perhaps this does also reflect a significant difference of aim. What I take to be the most important moral of such a conclusion is that it should encourage us to bring to bear the skeptical insights that have emerged from the recent history and sociology of science on contemporary science. But unlike, at least, the sociology of knowledge movement, I do not want to treat all parts of that science equally. To criticize everything is, in effect, to criticize nothing. It is precisely the importance of recognizing the disunity of science that it encourages us to try to sort the scientific sheep from the goats. To relativists and anarchists alike this will sound too close to the bad old normative philosophy of science. In mitigation I stress only that this is not to be a division in terms of the true scientific method, but one in terms of social worth. If there is one conclusion of overriding importance to be drawn from the increasing realization in recent times that science is a human product, it is that, like other human products, the only way it can ultimately be evaluated is in terms of whether it contributes to the thriving of the sentient beings in this universe.

Notes

Bibliography

Sources

Index

Notes

Introduction

1. This might be seen as a kind of grand-scale bootstrapping in (roughly) the sense of Glymour (1980).
2. See Dawkins (1986).
3. There is an irony here in the fact that the term *computer* originated as referring to people who made a living by calculating. Homunculus theories appear to be deeply embedded in the language with which these issues are currently addressed.
4. This term was introduced in Dupré (1981).
5. Quoted by Holton (1973), p. 366. Holton also recounts the story that Einstein, on being asked why he insisted on using ordinary hand soap for shaving, replied: "Two soaps? That is too complicated!" For the importance to Einstein of unification, see also Galison (1987), pp. 50–51.
6. Neurath is probably best known to contemporary philosophy through Quine's references to his figure of science as a boat undergoing repairs at sea (for example, 1960, p. 3). Thus his general antifoundationalism should be no surprise.
7. Cat, Chang, and Cartwright (1991) draw a close parallel between the views of Neurath and those of contemporary antireductionists such as myself (characterized as "postmodernists"). The main contrast, they suggest, is that whereas in the interwar period it seemed natural for progressive intellectuals to insist on the superiority of scientific authority to the mysticism, superstition, and incipient fascism that pervaded the Europe of the 1920s and 1930s, today science is widely perceived as co-opted to the interests of the ruling class.
8. An anonymous reviewer for Harvard University Press remarked that little written before 1962 was cited in this book. This is true, although

it does not reflect the view that the latest is necessarily the best. Rather, an attempt to trace the issues back another philosophical generation would have added pages and subtracted clarity.

9. This last idea is developed forcefully by Longino (1990).

10. An idea to which I am more sympathetic is that we should be very skeptical about the truth of the theories, but more optimistic about the existence of the entities (Cartwright, 1983; Hacking, 1983). A similar severing of the connection between the truth of theories and the existence of entities is a goal of the views of Putnam (1975), discussed in Chapter 1.

11. As far as I know, skepticism about science is no part of van Fraassen's purpose.

12. Examples are the current move to "naturalize" epistemology by appeal to so-called cognitive science; the project of replacing philosophy of mind and apparently the mental with neurobiology (discussed in Chapter 7); and in ethics, the idea that the speculations of sociobiologists might tell us something about how we should live (Ruse and Wilson, 1986).

1. Natural Kinds

1. The idea was apparently first proposed by Ghiselin (1966) but has been mainly championed by Hull (1989), pt. III.

2. See Putnam (1975), especially "The Meaning of 'Meaning'" (hereafter MM), "Is Semantics Possible?" (SP), and "Explanation and Reference" (ER); and Kripke (1972).

3. Locke (1689/1975), bk. III, chap. 3, sec. 15, p. 417.

4. Ibid., bk. II, chap. 23, sec. 12, p. 303. Locke shows himself well aware of the likely practical disadvantages of such equipment.

5. Ibid., bk. III, chap. 6, sec. 8, pp. 443–444.

6. The term *natural kind* has been used in rather different ways. Quine (1969), for instance, has used it in making the point that there are empirically discoverable distinctions in our "subjective quality space." These kinds, however, depend on the particular nature of human observers, and not necessarily on significant properties of the objects. They might, perhaps, better be referred to as "innate nominal kinds."

7. I use the term *nominal essence* here very broadly to include definitions, criteria, clusters of symptoms, and so on. I do not mean to imply that every kind requires an (even nominally) essential property.

8. See Achinstein (1968), pp. 91–105, 179–189; Hesse (1980), chap. 3.

9. MM, p. 269.
10. MM, p. 250; ER, p.204.
11. MM, pp. 227–229.
12. MM, pp. 229–234.
13. Presumably Putnam cannot mean to deny that even experts arrive at the reference of a term by way of some kind of Fregean sense. The difference must be that the sense (for experts, sometimes) is a description of the real essence. Denying the existence of the latter can leave only a difference of degree between the understanding of experts and of nonexperts. As Mellor (1977) observes, the division of linguistic labor provides no reason for abandoning a Fregean theory.
14. Kripke (1972) offers a distinct argument, one based on the necessity of self-identity. This amounts to the claim that if a kind is identical with the extension of some predicate, then this must be so necessarily, and the predicate must therefore name an essential property. Since the thrust of my argument is that there are no such properties and hence no such predicates, I take it that the presupposition of this argument will be sufficiently undermined.
15. For further critical discussion of Putnam's views, see Mellor (1977) and Zymach (1976).
16. If it is objected that the concept of a species was very different when European botanists first reached America, I will make the modestly counterfactual assumption that America was first discovered by Europeans in the late twentieth century.
17. Indeed, if there is really *no* other difference, it is impossible to conceive of any ground there could be for postulating this difference. This may be seen as an example of what Schlesinger (1963) has called the Principle of Connectivity.
18. Much of this importance may be attributable to the fact that Putnam is, or once was, a reductionist. A classic statement of a reductionist philosophy of science is that of Oppenheim and Putnam (1958).
19. For some purposes divisions into subspecies or varieties are required, but here these can safely be ignored. The frequent difficulty of deciding whether such a subspecific division should be taken as fully specific can be seen to reinforce the general pluralism of the present discussion.
20. All the preceding examples may be found in Spellenberg (1979).
21. See Borror and White (1970), pp. 218 ff.
22. This figure, taken from Borror and White (1970), has no doubt increased substantially in the past two decades.
23. See Putnam, MM, pp. 226–227.

24. An excellent account of the recent history of taxonomic theory can be found in Hull (1989). The issues will be taken up in Chapter 2.

2. Species

1. P. M. Churchland (1981, 1985); P. S. Churchland (1986). This view is discussed in Chapter 7.
2. Although the increasing dominance of phylogenetic conceptions of taxonomy has made this point more controversial. A more equal treatment of species and higher taxa is advocated by Ereshefsky (1991).
3. The main proponents of this thesis have been Ghiselin (1966, 1974) and Hull (1976, 1989), the latter having provided the most systematic and philosophically sophisticated defense of the position. A substantial number of philosophers and biologists have more recently accepted the thesis.
4. See Kitcher (1984a, 1984b) and Sober (1984b) for extensive discussion. Kitcher, in my view, makes the issue unduly complex by suggesting that species are *sets,* raising difficulties about how the membership of a set can change over time. Sticking with more traditional notions of *kinds* precludes these difficulties.
5. Those who believe that water has an essence (e.g., being constituted of H_2O molecules) may see a disanalogy between these cases. However, I do not see any reason in principle why a species defined, say, in terms of a large but indefinite cluster of properties could not satisfy the present argument. On most conceptions of species, if *enough* properties change we will have a new species.
6. See, for example, Hull (1989), pp. 82–83; Smart (1963).
7. The belief that there must be exceptionless laws true of the members of a natural kind is closely connected with the idea that to constitute a natural kind a set of objects must share a common essence. Laws governing the members of natural kinds might then be consequences of the common essence. This set of issues is addressed in Chapter 3.
8. As Hull (1989, p. 121) remarks, rather than put a bird and a butterfly in the same "niche" biologists are likely to add to the definition of the niche enough to keep the two kinds apart. I assume for purposes of this discussion that we are considering only sets of organisms superficially similar enough for the claim that they constitute a species to be taken seriously.
9. An alternative solution might be that species should be treated as an ontological category *sui generis* (see, for example, Mayr, 1987). This

seems to me to make the issue unnecessarily perplexing in view of availability of the pluralistic option I advocate.

10. Ghiselin (1987) and Mayr (1987) both accept that there are no species of such organisms; Mayr refers to them as "paraspecies," and Ghiselin as "pseudospecies"; for others (for example, Cowan, 1971), the existence of asexual species constitutes grounds for rejecting the biological species concept. Within bacterial taxonomy, which is Cowan's primary concern, pluralism has long been an attractive option.

11. The bible of this movement is Hennig's (1966) classic work.

12. Throughout this discussion I am assuming that whatever the epistemological problems with identifying genealogical branches, there is no problem in understanding what it is to be such a branch. This may not always be the case, however, as has been argued by Kitcher (1989), pp. 202–204.

13. Kitcher (1989), pp. 197–198, describes a case in which, according to the criterion of reproductive isolation, questions of conspecificity will depend on features of the environment.

14. For a more detailed discussion, see Mayr (1970), chap. 2.

15. Kitts and Kitts (1979) argue that there must be such a property, while admitting that they have no idea what it might be.

16. It is so supposed by Putnam (1975), p. 141.

17. Some empirical evidence on the extent of genetic variation in natural populations is presented by Lewontin (1974), chap. 3.

18. See Mayr (1970), p. 133. Not surprisingly, this is particularly likely in the case of sibling species, groups of morphologically almost identical species that are nevertheless reproductively isolated from one another (see Mayr, 1970, pp. 22–36).

19. This relates to a general difficulty with the phylogenetic species concept, that it may make correct assignments of organisms to species empirically inaccessible. It has, however, become very unfashionable to raise difficulties of this kind, which are liable to engender suspicions of verificationism or worse.

3. Essences

1. This claim is, of course, controversial in the light of the well-known views to the contrary of Kripke (1972) and Putnam (1975), discussed and criticized in Chapter 1. The possibility of deriving essentialism from semantic considerations has also been attacked by Salmon (1981).

2. Of course the relevant explanation here is one in terms of evolution,

or more specifically, common descent. In the present context, however, the question at issue is proximate causation not ultimate causation. Common descent is, on the other hand, one of the main
explanations of the existence of extraordinarily homogeneous classes
that do *not* share a common real essence.

3. More strictly, whether there are more sexual or asexual kinds of
organisms depends on how one counts. If, on the basis of a genealogical account of biological kinds, one considers every clone of an
asexual kind to be a separate kind, then there are many more kinds
of asexual organisms (see Hull, 1989, pp. 108–109).

4. Popper (1974) emphasizes that the same is true of most chemical
kinds, since heavy elements owe their existence to a particular cosmological history.

5. The nature of this pressure, however, remains surprisingly obscure.
Good sources on the problem are Williams (1975) and Maynard-
Smith (1978).

6. The classic text of sociobiology is Wilson (1975); an engaging popular
introduction is Dawkins (1976). Sociobiology has also been the subject of a good deal of criticism. The parts of the enterprise specifically
addressed to human nature have come under devastating attack from
Lewontin, Rose, and Kamin (1984), esp. chap. 9; and, perhaps a little
more sympathetically, Kitcher (1985).

7. The internal weaknesses of the sociobiological argument I have been
discussing are brought out clearly by Kitcher (1985), pp. 166–176.

8. The viability of this proposal will, of course, depend on accepting
that some taxonomic groupings are natural kinds. Someone who
holds the view that species are individuals would certainly not want
to suggest that such individuals might be formed by the union of
two kinds.

9. More recently, some feminists such as Jaggar (1983, p. 112) have
resisted such a sharp distinction, on the grounds that it suggests
erroneously that the sexual side of the dichotomy is absolutely rigid;
whereas, in fact, there is a continuous dialectical interaction between
cultural and biological aspects of gender differentiation. Relaxation
of the gendered distinction between physical activity levels of men
and women, for example, are said to have steadily reduced the
differences in size and strength between men and women. Holmstrom (1982) develops a similar position and defends the conception
of a distinctively female nature, on the basis that "nature" should be
understood in a way that encompasses both biologically and culturally determined aspects, since she also denies that these can be intel-

ligibly disentangled. While happily disavowing any implication that there is some readily distinguishable set of immutable biological differences between men and women, I believe that the distinction in the present context is useful and harmless. Like the vast majority of distinctions, it is neither sharp nor absolute. This limitation calls for caution rather than abstinence.

10. Some secondary sexual characteristics—such as the distribution of body hair—in reality show considerable geographic variation. If Darwin was right in attributing the majority of geographic variation among humans to sexual selection (Darwin, 1871, esp. chaps. 7, 19, 20), this fact is unsurprising.

11. See, e.g., Fausto-Sterling (1985); Longino (1990), chaps. 6 and 7. I shall return to the question of ideological bias in science in Chapter 11.

12. The relevant literature here is extremely large. Ortner and Whitehead (1981) provide a good illustrative collection of anthropological material. D'Emilio and Freedman (1988) trace the vicissitudes of a central aspect of gendered behavior, sexuality, through three and a half centuries of American history.

13. These are favored objects of study in sociobiological work on rape (see Barash, 1979; Thornhill and Thornhill, 1983). The conclusions about ducks, however, are open to serious question (see Fausto-Sterling, 1985, pp. 191–192).

14. A detailed and somewhat technical treatment of this topic has been provided by Boyd and Richerson (1985).

15. This idea is developed in more detail in Dupré (1987b).

16. Hrdy (1981) offers a feminist version of parts of sociobiology. At a more popular level, Morgan (1972) provides an entertaining feminist answer to Desmond Morris.

17. Jaggar (1983) suggests that a commitment to biological determinism grounded in human sexes with distinct essences as the explanation of gender differences is a characteristic defect of radical feminist thought (the strand of feminism that treats gender oppression as *the* most basic political problem).

18. I explore this problem in Dupré (1990a).

19. Such as that of Goldberg (1973). Unfortunately, it is my impression that some feminists have been led by the same observation in the same direction, although certainly those who, like Goldberg, see male aggressiveness as the crucial, and even biologically determined, factor are likely to point out that aggressiveness is not necessarily an unqualified virtue.

20. Part of the explanation of male supremacy, presumably, is in terms of cultural evolution. Male supremacy is undoubtedly buttressed by institutions and beliefs of varying degrees of antiquity, including, for instance, more or less central aspects of most major religions. It is even conceivable that a combination of such cultural factors might unite to explain *all* male supremacy as a true cultural evolutionary homology, deriving from some primeval ancestral male-dominated society. Perhaps genuinely biological forces serve to make such cultural factors particularly recalcitrant. There are, at any rate, explanations along these (and no doubt other) lines that do not at all encourage the conclusion that male supremacy is immutable.

21. Such a potentially interminable narrowing of categories is actually implicit in various contemporary efforts to characterize laws of nature. This issue will be taken up in detail in Part III.

22. Many females, that is to say, supply all the energy and work that are required to raise their young, and yet transmit only half their genes to the next generation. It is mysterious that they are not rapidly supplanted by parthenogenetic females capable of transmitting twice the genes for the same amount of effort.

23. Atoms are often suggested as an example of natural kinds, with atomic number serving as an essential property. But the fate of cars driven over salted roads for any time provides a reminder that iron atoms are not at all the same as ferric ions, although both have atomic number 26. Atoms are also said to vary with respect to transitory states of orbital electrons, properties said to be of great significance to their chemical behavior. (See further Zymach, 1976; Dupré, 1983, pp. 326–327.)

4. Reductionism and Materialism

1. A classic statement is by Oppenheim and Putnam (1958), from whom much of the subsequent account has been derived. The position is discussed extensively, and with considerable sophistication, by Nagel (1961). More recently, technical details have been worked out painstakingly by Causey (1977). My terminology is perhaps confusing, in that it may reinforce the common supposition that associates this kind of reductionism with classical positivism. Certainly it does not seem to be what Neurath meant by the "unity of science" (see Cat, Chang, and Cartwright, 1991). Nevertheless, it seems so generally to be the view that philosophers now have in mind when they speak

casually of reductionism, that I think it has fully earned the designation "classical."

2. It is, of course, entirely possible that there should be no lowest level. There might, for example, be an infinite hierarchy of levels, or at some level we might discover homogeneous and infinitely divisible physical stuff. In the first case we would have to accept some particular level somewhat arbitrarily as the starting point of the reductionist hierarchy; in the second we would presumably aim at an account of the production of a lowest level of discrete entities as a consequence of whatever laws govern the behavior of physical stuff.

3. Feinberg (1966) and Smart (1978) are both prepared to commit physicalism to something close to our contemporary physical theories. Healey (1978–79) has provided a valuable discussion of this and various issues concerning contemporary materialism or physicalism.

4. A similar idea is described by Hellman and Thompson (1975) as the "Principle of Physical Exhaustion."

5. See, e.g., Quine (1960), esp. chap. 7.

6. Since a materialist must, obviously, believe in the existence of physical entities, and since the only ground for such a belief is, presumably, its contribution to the explanation of the behavior of observable phenomena, there is a kind of interdependence between materialism and this view of ontology.

7. Later in this chapter I shall, however, consider an argument that appears to lead from compositional materialism to some kind of reductionism, and hence to monism.

8. This distinction is discussed in greater detail by Nickles (1973).

9. See, e.g., Ruse (1971).

10. Elsewhere I have referred to this idea as "theological reductionism" (Dupré, 1983).

11. A possibility, incidentally, mentioned by Duhem (1906/1954), pp. 139–141.

12. Moore did not use the term *supervenience,* but, as Kim (1978) notes, he clearly stated the basic idea (Moore, 1922, p. 261). The term is used by Hare (1952), pp. 80–81.

13. For example, Davidson (1980), p. 214.

14. See Rosenberg (1985). Collier (1988), on the other hand, argues that supervenience is a concept poorly suited to expressing the relations between levels of the biological hierarchy.

15. For further discussion of the relation of supervenience theses to reductionism, see Kim (1978) and Healey (1978–79).

16. See Putnam (1981); Burge (1986). The idea derives most generally

from the perspective on language and the mental developed by Wittgenstein (1953). It would be possible to reply that the supervenientist does not have to limit the set of facts supervened on to the physical state of the particular subject of the mental state in question. This, however, is the beginning of a slippery slope. If supervenience is allowed to extend indefinitely in time and space, it entirely escapes the reach of empirical evidence. In that case it will also lack any grounds of credibility.

17. As early as 1956, Kemeny and Oppenheim argued in an influential paper that the crucial feature of a reduction was not that it should provide a derivation of the reduced theory, but only that the observational consequences of the reduced science should be derivable from the reducing science.

18. Although some contemporary theories of the replacement of the mental by the neurophysiological seem to deny this point (e.g., P. S. Churchland, 1986).

19. Influential arguments closely related to this one, in connection with the physicality of the mental, will be discussed in Chapter 7.

20. It does not matter whether this is spelled out in terms of propensities or in terms of long-run frequencies, although perhaps radically subjectivist views of probability would be difficult to reconcile with the idea of causal completeness.

21. On this point see also Duhem (1906/1954), pp. 23–27.

22. Such an interpretation is suggested by the influential thesis (criticized in Chapter 1) that natural kind terms rigidly designate an extension that is thus independent of changes in the theory of the objects concerned (though perhaps dependent on some "ultimate" theory).

23. Wittgenstein (1953), p. 178. For an elegant recent statement of a related opinion, see Hampshire (1991).

5. Reductionism in Biology 1

1. The best-known introduction to the philosophy of biology (Hull, 1974), for example, devotes about 40 percent of the text to the topic of reductionism. Reduction in genetics, the main specific concern, will be addressed in the next chapter.

2. The general importance to biological science of this kind of model-building methodology is emphasized by partisans of the so-called semantic view of theories. See, e.g., Thompson (1988) or Lloyd (1988).

3. I assume for the sake of simplicity that the number of parasites per

parasitized host is 1. The generalization of this assumption is obvious and trivial.

4. For a more detailed discussion of this and a range of similar models, see May (1973).

5. My condensed exposition of this distinction is derived from Gendron (1970).

6. Illustrating the provisional nature of such explanatory projects, it has been suggested that this phenomenon is best explained by the fact that increases in these populations encourage increased activity by human hunters. If this is right, it would presumably be possible to construct some kind of model of the interactions between human hunters and furry animals. As will, I trust, become clear, my development of this example is not intended to be strictly factual.

7. For which see, e.g., Odum (1971), pp. 191–192.

8. In case this exposition should seem to conflict with the *realism* defended in Part I, I should point out that it is *descriptions* that are idealizations. The kinds that they imperfectly describe are real enough. Both the physiologist and the ecologist intend to increase our understanding of real hares. The classes of organisms to which their illumination extends may or may not be the same.

9. The general intuition involved here will be defended at greater length in Part III.

6. Reductionism in Biology 2

1. Contributions to this considerable literature generally supporting a reductionist interpretation include Schaffner (1967, 1976), Ruse (1973, 1976), and Goosens (1978); such positions have been exposed to powerful objections by Hull (1972, 1974), Wimsatt (1976, 1980), Kitcher (1982, 1984c), and, somewhat reluctantly, Rosenberg (1985). A recent attempt to rebut these criticisms has been offered by Waters (1990).

2. An important analysis of the philosophical significance of the rise of molecular genetics, also emphasizing the diversity of uses of the term *gene* in ways complementary to the discussion here, is that of Kitcher (1982).

3. One thing that is clearly established by our present knowledge of DNA is that the objects to be subjected to this classification are perfectly real.

4. At the present time, what is surely one of the most ambitious data collection enterprises in the history of science, the human genome

project, aims to provide the entire range of possible referents for genetic kind terms, at least for one species or for its (somehow) selected representatives.

5. See Hull (1974), pp. 18–19.

6. The most important of these are well distinguished and criticized by Waters (1990).

7. Waters (1990) does not explicitly address this dichotomy, although attention to it could help bring out the force of the argument of Kitcher (1984c) that he refers to, somewhat misleadingly, as the "Gory Details Argument." The point is not just that the molecular details of gene replication are appallingly complicated, but that we are interested in what the cell is *doing,* a question about function to which the answer may remain invariant across an extremely heterogeneous set of underlying molecular processes.

8. See Wright (1976); Millikan (1984). For a very different account, see Cummins (1975).

9. Rosenberg (1989) suggests that the prevalence of many-many relations between genes and traits is explained by the fact that only functions, not structures, are visible to natural selection. Although this seems very likely correct, I do wish to dissent from the implication that such a relation *requires* any such causal explanation. Consider, for instance, the relation between animals and omnivores. Many kinds of animals are omnivorous. But there are groups of people who are strictly vegetarian, and lineages of pigs that eat only the most homogeneous swill. Perhaps the reason a causal explanation does not seem appropriate in this case is that onmivory is a natural kind only relative to an interest in ecology.

10. Of course we *might* have special reasons to be concerned with chemical details, for example in investigating evolutionary relations on the basis of mutation rates. Needless to say, I am not concerned by the fact that different inquiries may require different classificatory principles.

11. For example, Putnam (1960, 1963); Fodor (1968).

12. One might possibly suggest, in the spirit of views on species discussed in Chapter 2, that human hemoglobin genes, for example, are not a kind at all, but a lineage. But it is surely clear that from the perspective of physiology hemoglobin is a kind. Thus the pluralism of ontological categories that I suggested as a consequence of the species-as-individuals hypothesis would follow even more clearly in this case.

13. An important complication that I shall not discuss is that many genes

do not code for any peptide chain, either because they serve some higher-order regulatory function (activating and deactivating other genes, for example) or because they serve no function at all (so-called junk DNA). I take it that explicit consideration of this problem could only reinforce my general thesis.

14. See Wilson (1975), p. 554.

15. I say only "likely" in view of the *ceteris paribus* clause (see the quotation from Dawkins on page 124).

16. Dawkins, however, is not entirely consistent in his adherence to his official conception of the meaning of "gene for X." For example, a few pages after the passage quoted above he writes: "to use 'a gene for X' as a convenient way of talking about 'the genetic basis of X' has been a standard practice in population genetics for over half a century" (1982, p. 29). This statement suggests that the gene *is* being identified with whatever part of the genetic material is causally relevant to the presence of X. Whether or not this criterion succeeds in determining any specific part of the DNA, the view suggested is clearly quite different from the one that Dawkins predominantly espouses.

17. Kitcher concludes a discussion of this topic with the provocative statement that "there is no molecular biology of the gene. There is only molecular biology of the genetic material" (1982, p. 357).

18. Here I differ from the main conclusion of Rosenberg (1989). As I understand his argument that the failure of reductionism implies antirealism about the entities referred to in biological theory, it is grounded on a more consequential physicalism than I wish to accept.

19. This is related to the molecular statics of which Putnam's well-known example of the square peg in the round hole is part (1978, pp. 42–43).

20. The deficiency I am claiming for selection coefficients may be seen as the lack of what Goodman (1965) described as "projectibility."

21. I do not propose here to go in great detail into the ramifications of the units-of-selection debate, a debate that has perhaps received more attention than any other issue from philosophers interested in biology, although I shall discuss one major position on the debate, genic selectionism, in the final section of this chapter. Here and in that discussion I advocate a pluralistic conception of natural selection, with most selection occurring at the individual level. But rather than attempting to defend this position in detailed engagement with the extensive literature on the question, I hope to derive its plausibility directly from the philosophical picture developed throughout this

book. (For references on the units-of-selection debate, see note 28, below.)

22. Except, of course, in those cases such as meiotic drive or so-called junk DNA, in which selection processes do, in just the ways typically true of organisms, act on genes. In these cases we should see the organisms, this time, being dragged along by processes happening at the lower level. The appropriateness of this description is confirmed by the fact that they may be taken to phenotypic states involving significant losses of fitness. The analogy can be completed by noting that selection on phenotypes can constrain such genetic processes, for example by lethality.

23. With the addition of a dimension representing overall fitness, this landscape of genetic possibility becomes what is sometimes referred to as an adaptive landscape. Since overall genetic fitness strikes me as a notion little more coherent than the fitnesses of individual genes, I do not intend the present metaphor to admit of this expansion.

24. As an extreme example of the difficulty with such a demand, Cohn (1987) gives a fascinating account of how learning the technical terminology of Pentagon and academic "defense" experts subtly hampered her ability to express her dissent from their insane fundamental premises.

25. This is as far as I can concur with the pluralism of models advocated by Maynard-Smith (1987) and Sterelny and Kitcher (1988).

26. As noted above, real benefits can be attributed to the collection of data on genetic features of populations, especially the refutation of "scientific" racism.

27. I say "little" rather than "nothing" in view of my earlier remarks about the (possible) illumination of certain questions about evolutionarily accessible states. But only insofar as this illumination is translatable into some general insight into evolutionary possibility described at the phenotypic level does this qualify as significantly contributing to the layperson's understanding of natural processes.

28. As mentioned in note 21 above, this issue has been widely discussed. The best general overviews, both favoring broadly pluralistic, though in other respects very different, positions, are by Sober (1984a) and Lloyd (1988). The thesis of genic selectionism has been most prominently advocated by the biologist Richard Dawkins (1976, 1982). Apart from the books by Sober and Lloyd just cited, influential critiques of genic selectionism are to be found in Sober and Lewontin (1982) and Brandon (1982). For debate subsequent to

the paper by Sterelny and Kitcher, see Sober (1990) and Kitcher, Sterelny, and Waters (1990).

29. A large proportion of this debate, starting with the influential discussion by Sober and Lewontin (1982) of some difficulties with representing heterozygote superiority in terms of selection on alleles, has concerned the question whether selection should be conceived in terms of alleles or genotypes, the former position being identified with the thesis of genic selectionism. Since my concern in this chapter is with the relation between genotypic and phenotypic conceptions of selection, the question about allelic versus genotypic selection is not of direct relevance to my argument.

30. Much of the criticism, especially that of Sober (1984a) and Brandon (1982), depends, though in rather different ways, on claims that genes do not have the right kind of causal relation to the organism-level interactions that constitute most selection processes. Although I disagree with the way these claims are substantiated (especially Sober's version, which will be discussed in detail in Part III), I am sympathetic to the general position.

31. The most detailed defense of this alternative position is that by Sober (1984a).

7. Reductionism and the Mental

1. Various responses to Churchland's book (mostly more sympathetic than the discussion here), together with replies by Churchland, can be found in *Inquiry* 29, no. 2 (1986). Subsequent page numbers in the text all refer to Churchland (1986).

2. Henceforward I shall not use qualifiers (such as "so-called") or scare quotes in connection with this expression. It should be understood that I use the term with strong reservations.

3. The important exceptions are those cases in which behavior is explained by gross physiological pathology. It is a common fallacy to suppose that the role of neurobiology in explaining pathological behavior provides an argument for its potential usefulness in explaining normal behavior—as if identifying a necessary condition for normal behavior, an intact cerebrum, allowed one to infer to the sufficiency of conditions of approximately the same kind.

4. The idea that science is, on the contrary, a complex human practice has become something of a commonplace in recent philosophy. Var-

ious versions of this insight can be traced to Kuhn's (1970) classic work. See, e.g., Rouse (1987), chap. 2.

5. The nearest Davidson comes to arguing explicitly for this claim in "Mental Events" is where he is discussing the laws that could underlie psychophysical interactions: "Physical theory promises to provide a comprehensive closed system guaranteed to yield a standardized, unique description of every physical event couched in a vocabulary amenable to law. It is not plausible that mental concepts alone can provide such a framework, simply because the mental does not, by our first principle, constitute a closed system" (1970, p. 99). This contrast lies at the heart of Davidson's position. One may well wonder whether the assumption that interaction with the physical prevents the mental from forming a closed system but not vice versa does not beg the question of the materiality of the mind. Some of Davidson's later arguments (1973) do seem more specifically directed at the absence of psychological laws.

6. Davidson's theory of causality is laid out in more detail in an earlier article (1967).

7. The classic source both for the elaboration of this point and for the exploration of how everyday causal statements might be made precise is Mackie (1974).

8. See also Hornsby (1980–81).

9. Thus I concur with philosophers such as Salmon (1984) who see causality as ultimately consisting of continuous processes rather than lawlike relations between events.

8. Determinism

1. This concept has been the subject of a good deal of technical discussion. However, I use it here with neither a rigorous definition nor an apology, since it seems to me also to be the appropriate term for the nontechnical concept I am considering.

2. A useful discussion of the history of this problem and its implications is provided by Suppes (1984), pp. 125–130.

3. One might take determinism to be neither an epistemological nor a metaphysical doctrine, but rather a characteristic of certain *theories*. This seems, for instance, to be the position of Nagel (1961), who notes that from the fact that classical mechanics is a deterministic theory, it does not follow that Laplace's demon can rely on a knowledge of mechanics to predict the future course of events, since many causally relevant properties (electromagnetic, thermal, chemical, and

so on) are not part of mechanics (pp. 282–283). This sense of determinism may ultimately be the most useful. It is, however, tangential to the main concerns of this book.

One further interpretation of determinism that should be mentioned is as a regulative methodological position. This will be discussed later in the chapter.

4. Lewis (1973), p. 559; cited in Rosen (1982–83), p. 103.

5. A trenchant and forthright critic of determinism is Rosen (1978, 1982–83). Crucial recent contributions to weakening the hegemony of determinism are Suppes (1970, 1984) and W. C. Salmon (1984; Salmon, Jeffrey, and Greeno, 1971).

6. By *irreducibly* indeterministic, I mean that the failure of indeterminism does not simply reflect our ignorance of some crucial parameter or parameters. An example of reducible indeterminism would be the truth of one of the so-called hidden variable theories often discussed in connection with quantum mechanics. The distinction between reducible and irreducible determinism will be discussed further later in the chapter.

7. For a useful discussion see Nagel (1961), pp. 312–316.

8. See, e.g., Hempel (1965), pp. 58–69; Nagel (1961), pp. 503–520.

9. Hempel (1965), pp. 381–412. The account I shall give here deviates in some important respects from Hempel's and bears affinities to that developed by Railton (1978), which he refers to as a "deductive nomological model of probabilistic explanation."

10. This should be read: "the probability of E (an event of the effect type) at time $t+1$, given the occurrence of C (an event of the cause type) at time t."

11. For more-detailed discussion, see Jeffrey and Salmon in Salmon, Jeffrey, and Greeno (1971); and Salmon (1984), chap. 2.

12. High probability is not a requirement of Railton's (1978) account. In Railton's model the conclusion of a probabilistic explanation is some probability statement, supplemented by a "parenthetic addendum" to the effect that the event in question in fact occurred. This can be seen to provide an explanation only of the *probability* of an event, not of its occurrence. (If it were intended to provide the latter, the parenthetical addendum would presumably make it viciously circular.) Although interesting issues are raised about what are the most appropriate objects of explanation in an indeterministic context, this divergence from the issue of explanation of events renders Railton's account tangential to my present concerns.

13. Such a move was advocated by Salmon and Jeffrey (Salmon, Jeffrey,

and Greeno, 1971) and by Suppes (1970), who worked out the details
of the alternative conception in considerable detail.

14. This example is drawn from Salmon (Salmon, Jeffrey, and Greeno,
    1971), p. 34. Note that it is not a problem for the S-R view, since
    taking birth control pills does not affect the probability of pregnancy
    for men.

15. See Mackie (1974), p. 44.

16. In Hempel's terminology, we have met the "requirement of maximal
    specificity" (1965, pp. 399–400).

17. These I take to provide the ontological grounding of causal relevance
    laws, a matter taken up in more detail in the next chapter.

18. Earman (1986) discusses a variety of examples from physics in con-
    siderable depth. I shall not attempt to evaluate or expound his con-
    clusions. One outcome of his analyses is that it is a surprisingly
    difficult question whether a particular system of scientific laws is
    deterministic or not. As far as I can see, his conclusions do not affect
    the more global questions I am now considering.

19. A point familiar to anyone who has ever watched a baseball game at
    Wrigley Field or Candlestick Park. As this example may also remind
    us, some falling projectiles—even wind-blown ones—are interfered
    with by autonomous human agents.

9. Probabilistic Causality

1. There is a large literature developing this idea. Some basic texts,
   mentioned in the previous chapter, are Suppes (1970) and Salmon,
   Jeffrey, and Greeno (1971). Important more recent discussions include Cartwright (1979; reprinted in Cartwright, 1983), Salmon
   (1984), and Eells and Sober (1983).

2. Of course, this explanation might very well itself reflect more subtle
   and deeply rooted biases.

3. One other author who has criticized the unanimity condition is
   Gifford (1986).

4. Eells (1987) responds to arguments originally proposed in Dupré
   (1984). I reply to Eells's criticisms in Dupré (1990b). This chapter
   recapitulates substantial parts of the exchange.

5. I say "generally" because I am not sure that this is a universal truth
   about probabilistic causal relations. Quantum mechanics, assuming
   it contains probabilistic and causal laws, is normally taken to apply
   to whole kinds (electrons, fermions, and so on) rather than to particular populations. If I am right in arguing that humans, birds,

sulfur-bellied flycatchers, and so on can sometimes be understood as kinds, the relativization to populations is a much more contingent aspect of probabilistic laws. I am inclined to think that "Smoking causes heart disease" may very well be a law that applies to humankind.

6. As implied above, it cannot be assumed that all these sets of conditions will actually be realized in a given population. In cases in which they are not, the defender of unanimity could either declare the causal truth about the population to be epistemically inaccessible or decide that only exemplified sets of conditions mattered. The latter would have some paradoxical consequences. For example, it might happen that the birth of a baby with a particular unique physical constitution would make it cease to be true that smoking causes heart disease. Neither solution looks encouraging for the unanimity theorist.

7. I shall ignore the familiar issues about placebo effects and the like, which are crucial to the application of this methodology to human questions. The theoretical point they illustrate is just that we must be sure that the cause is not introduced in a way that simultaneously carries with it some other causally efficacious factor (such as the belief that one has received an effective treatment).

8. See also Giere (1984), p. 284.

9. As the discussion in Chapter 7 makes clear, I do not intend this suggestion to privilege scientific investigation over commonsense knowledge in every case.

10. Suppose no smokers exercise. We must then compare samples both of which contain no exercisers. There is no immediate problem since smoking and exercising cannot interact. However, it would be necessary to investigate such possibilities as that smoking caused heart attacks by preventing exercising.

   As Eells obliquely suggests (1987, p. 111, n. 3), this proposal has much in common with the theory of Giere (1984). One difference is that Giere, in discussing spurious correlations, suggests (pp. 293, 301) that we adjust the frequency of other factors in a control sample to match their frequencies in a test sample. My proposal is rather to match the frequency of other factors in the test sample to their frequency in the general population. This seems preferable in principle because surely we are ultimately interested in the impact of a factor on the actual population with its actual distribution of other factors.

   I am grateful to Eells (1987) for helpful clarification of the proposal roughly mooted in my 1984 article.

11. This inversion of the Humean picture has been advocated by a small minority of philosophers in the post-Humean history of the topic. Notable recent examples are Ducasse (1924) and Harré and Madden (1975). Nancy Cartwright and I have developed in rather greater detail elsewhere (Dupré and Cartwright, 1988) one of the central arguments about to be discussed for drawing this conclusion from the belief in a probabilistic metaphysics of causality. Cartwright (1990) has more recently provided an extended exposition of the notion of a causal capacity. I shall not attempt to provide an analysis of this notion. The rough idea is that a capacity is a kind of property, one that is associated with either a property (*smoking* causes heart disease) or a kind of thing (*aspirins* cure headaches). Although this account indicates a kind of generality, it is not a generality that carries any expectation of indefinite refinability of the kind generally advocated by theorists of natural laws.

12. Compare Smart's (1963) similar claim about biological laws.

13. For a discussion complementary to the preceding, see Suppes (1984), pp. 55–59.

14. Throughout this discussion I mean by "opposed capacities" the simultaneous capacity to produce and to prevent a particular effect. Some weaker kinds of dual capacities are distinguished by Dupré and Cartwright (1988, pp. 525–526). The present argument is developed in more detail in that article.

15. Different causal intermediaries leading to opposite effects provides one class of cases of opposed capacities (a much-discussed case was introduced by Hesslow, 1976). But I see no reason why opposed effects should not be produced directly by the same cause (see further Dupré and Cartwright, 1988). However, bearing in mind the doubts about causal chains raised in Chapter 7, I am uncertain whether this is ultimately a meaningful distinction.

16. Eells (1988) does attempt to address this issue by relativization of such causes to limited subpopulations. This attempt is criticized by Dupré and Cartwright (1988, pp. 528–529).

17. The most imaginative statistical inquiries of this kind of which I am aware can be found in the yearly editions of the *Bill James Baseball Abstract* (New York: Random House).

18. For that matter, different *kinds* of pitchers—power pitchers, junkballers, sinker/slider pitchers, knuckleballers, etc.—have different characteristic capacities to get batters out in particular ways (strikeouts, ground balls, etc.) and characteristic tendencies to give up hits (occasional home runs, fairly frequent ground ball singles, etc.).

19. This kind of agnosticism is remarkably rare in the philosophical literature on causality. An exception is an interesting paper by James Woodward (forthcoming).

20. The centrality of more or less long-term plans, as opposed to more transient intentions, in the analysis of human action has been appropriately emphasized by Bratman (1987).

21. I am inclined to think this is generally overstated. Certainly in the broadest metaphysical sense that I have been advocating, that of constituting rich depositories of causal capacities, other animals have much in common with humans. I have argued elsewhere against the alleged human monopoly on mentality (Dupré, 1990c).

10. The Disunity of Science

1. A phenomenon often referred to as physics envy. (I do not know who was first responsible for this joke.) Perhaps the most dramatic illustration of this phenomenon is the apparent willingness of a country plagued by homelessness, hunger, massively inadequate access to medical care, and a collapsing educational system to spend tens of billions of dollars on machines (such as accelerators) the operation of which not one person in ten thousand understands, and the purpose of which, arguably, no one understands. The irresistible parallel is with the erection of Gothic cathedrals, although these, at least, had outstanding aesthetic virtues.

2. Ironically, the "equivalence postulates" advocated by sociologists of knowledge (e.g., Barnes and Bloor, 1982) appear to satisfy this demand. I shall discuss this idea briefly in the next chapter.

3. Hacking (forthcoming) distinguishes a number of quite distinct senses of scientific unity.

4. Darden and Maull note parallels between this conception and Lakatos' (1978) "research programmes," Kuhn's (1970) "paradigms," and especially Toulmin's (1972) "disciplines." It is also relevant to my remarks in Chapter 7 concerning the complexity of scientific practice as an objection to eliminative reductionism.

5. The conclusion of their paper is explicitly agnostic about the scope of their analysis. Maull (1977) may be slightly more sympathetic to a broader interpretation but is not committed to any strong unificatory hypothesis.

6. See, e.g., Galison (1987).

7. The case is discussed, though in a somewhat different context, by Hempel (1966, pp. 52–54).

8. This echoes Hull's treatment of species as individuals, discussed in Chapter 2.

9. Some further reasons are discussed in Dupré (1990d). Another way of stating my view is that I am unfashionably resistant to the total naturalization of epistemology.

10. Nancy Cartwright (1983) makes the same claim for physical theories of broad scope. If she is right, the distinction I have in mind is only a matter of degree. It may still be a significant one. Repeating the skeptical note that has sounded from time to time in the text, I suggest that these formalized but empirically questionable sciences may have very tenuous scientific credentials.

11. Thus although, as I shall explain further in the next chapter, I greatly admire the contributions of Paul Feyerabend (e.g., 1975) to our understanding of science, my epistemological pluralism stops short of the anarchism he advocates.

12. The importance of this virtue is argued persuasively and at length by Longino (1990), esp. chap. 4.

11. Science and Values

1. For background on this topic, see Achinstein (1968); Hesse (1980), chap. 3.

2. Hobbes, quoted in Shapin and Schaffer (1985), p. 128.

3. The elitist and hierarchical structure of contemporary high-energy physics, the modes of ascent through the hierarchy, and the complex space constructed around the vast experimental machines of this discipline are fascinatingly explored by Traweek (1988).

4. Quoted in Shapin and Schaffer (1985), p. 128.

5. This is an important part of the background to the argument by Feyerabend that in a genuinely democratic society, science would have to be under the control of educated laypeople (1978, pp. 88–91 and 96–98). Feyerabend's views will be discussed further at the end of this chapter.

6. For further discussion, see Birke (1986), pp. 22–25.

7. These assumptions are greatly reinforced by the role of evolutionary adaptationism in contemporary biology and related human disciplines, mediated by the observation that some degree of heterosexual activity is a precondition for reproductive success. But even the hypothesis that there may be a genetic basis for heterosexual behavior (whatever such a hypothesis is intended to imply) does nothing to license the treatment of homosexuals or heterosexuals as *kinds*. There

is, after all, a genetic basis for having blue eyes or webbed toes, but nobody takes this to imply that people with these features form kinds deserving systematic investigation.

8. An outstanding survey of scientific mistreatments of men and women as biological categories has been provided by Fausto-Sterling (1985). Rhode (1990) offers an excellent and uniquely interdisciplinary survey of current feminist approaches to the question of sexual difference.

9. See Birke (1986), chap. 4.

10. I have developed detailed suggestions roughly along these lines elsewhere (Dupré, 1987b, 1990a).

11. Haraway (1981–82, 1985) has provided a radical and important account of the way in which late twentieth-century science has become a site for the creation of social order. She identifies the central paradigm of much contemporary science as information science, which, in turn, she traces to logistical techniques developed during World War II. True to these origins, the central problematic of this paradigm is one of command and control. Haraway traces the infiltration of much of biology by this paradigm during the postwar years, more specifically, by "the translation of the world into a problem in coding" (1985, p. 83). This transmutation is illustrated by the development of molecular genetics, ecology, sociobiology, and immunobiology. According to Haraway, the objects that are studied by these disciplines are in large part constructed to fit the needs of the informational networks into which they are to be fitted, a construction indicated by the term *cyborg*—part cybernetic, part organism—which applies (among other things) to humans located in such informational systems. The construction of the objects of science has been explored further in *Primate Visions* (1989), Haraway's brilliant account of the history of primatology. Despite having much relevance to my present aims, Haraway's work is not easy to integrate with my very different approach, so I regretfully consign it to this note.

Bibliography

Achinstein, P. 1968. *Concepts of Science*. Baltimore: Johns Hopkins Press.

Barash, D. 1979. *The Whisperings Within*. New York: Harper & Row.

Barnes, B., and D. Bloor. 1982. "Relativism, Rationalism, and the Sociology of Knowledge." In *Rationality and Relativism,* ed. M. Hollis and S. Lukes. Cambridge, Mass.: MIT Press.

Bewick, T. 1824. *A General History of Quadrupeds*. 8th ed. London: Longmans.

Birke, L. 1986. *Women, Feminism, and Biology*. Brighton: Harvester Press.

Borror, D., and R. White. 1970. *A Field Guide to the Insects of North America North of Mexico*. Boston: Houghton Mifflin.

Boyd, R., and P. Richerson. 1985. *Culture and the Evolutionary Process*. Chicago: University of Chicago Press.

Brandon, R. N. 1982. "Levels of Selection." In *PSA 1982,* ed. P. Asquith and T. Nickles. East Lansing, Mich.: Philosophy of Science Association.

Bratman, M. 1987. *Intentions, Plans, and Practical Reasons*. Cambridge, Mass.: Harvard University Press.

Bruce, J. 1989. "Homes Divided." *World Development* 17:979–991.

Burge, T. 1986. "Individualism and Psychology." *Philosophical Review* 95:3–45.

Carnap, R. 1928. *Der logische Aufbau der Welt*. Berlin: Weltkreis. Translated by R. A. George as *The Logical Structure of the World*. London: Routledge and Kegan Paul, 1967.

Cartwright, N. 1979. "Causal Laws and Effective Strategies." *Noûs* 13:419–437.

——— 1983. *How the Laws of Physics Lie*. Oxford: Oxford University Press.

——— 1990. *Nature's Capacities and Their Measurement*. Oxford: Oxford University Press.

Cat, J., H. Chang, and N. Cartwright. 1991. "Otto Neurath: Unification as the Way to Socialism." In *Einheit der Wissenschaften*, ed. J. Mittelstrass. Berlin: Akademie der Wissenschaften.

——— Forthcoming. "Political Philosophy of Science: Otto Neurath, Unity of Science, and Socialism." In *The Disunity of Science*, ed. P. Galison and D. Stump. Stanford: Stanford University Press.

Causey, R. L. 1977. *The Unity of Science*. Dordrecht: D. Reidel.

Churchland, P. M. 1981. "Eliminative Materialism and the Propositional Attitudes." *Journal of Philosophy* 78:67–90.

——— 1985. "Reduction, Qualia, and the Direct Introspection of Brain States." *Journal of Philosophy* 82:8–28.

Churchland, P. S. 1986. *Neurophilosophy*. Cambridge, Mass.: Bradford Books/MIT Press.

Cohn, C. 1987. "Sex and Death in the Rational World of Defense Intellectuals." *Signs: Journal of Women in Culture and Society* 12:687–718.

Collier, J. 1988. "Supervenience and Reduction in Biological Hierarchies." *Canadian Journal of Philosophy*, suppl. vol. 14:209–234.

Cowan, S. T. 1971. "Sense and Nonsense in Bacterial Taxonomy." *Journal of General Microbiology* 67:1–8.

Cracraft, J. 1983. "Species Concepts and Speciation Analysis." *Current Ornithology* 1:159–187.

——— 1987. "Species Concepts and the Ontogeny of Evolution." *Biology and Philosophy* 2:329–346.

Cummins, R. 1975. "Functional Analysis." *Journal of Philosophy* 72:741–765.

Darden, L., and N. L. Maull. 1977. "Interfield Theories." *Philosophy of Science* 44:43–64.

Darwin, C. 1871. *The Descent of Man, and Selection in Relation to Sex*. Reprint, Princeton: Princeton University Press, 1981.

Davidson, D. 1967. "Causal Relations." *Journal of Philosophy* 64:691–703. Reprinted in Davidson, 1980.

——— 1970. "Mental Events." In *Experience and Theory*, ed. L. Foster and J. Swanson. Amherst: University of Massachusetts Press. Reprinted in Davidson, 1980.

——— 1973. "The Material Mind." In *Logic, Methodology, and Philosophy of Science*, ed. P. Suppes et al. Dordrecht: North Holland. Reprinted in Davidson, 1980.

——— 1980. *Essays on Actions and Events*. Oxford: Oxford University Press.

Dawkins, R. 1976. *The Selfish Gene.* Oxford: Oxford University Press.

——— 1982. *The Extended Phenotype.* Oxford: Oxford University Press.

——— 1986. *The Blind Watchmaker.* New York: W. W. Norton.

D'Emilio, J., and E. B. Freedman. 1988. *Intimate Matters.* New York: Harper & Row.

Donoghue, M. J. 1985. "A Critique of the Biological Species Concept and Recommendations for a Biological Alternative." *Bryologist* 88:172–181.

Ducasse, C. J. 1924. *Causation and the Types of Necessity.* Seattle: University of Washington Press.

Duhem, P. 1906/1954. *The Aim and Structure of Physical Theory,* trans. P. P. Weiner. Princeton: Princeton University Press.

Dupré, J. 1981. "Natural Kinds and Biological Taxa." *Philosophical Review* 90:66–90.

——— 1983. "The Disunity of Science." *Mind* 92:321–346.

——— 1984. "Probabilistic Causality Emancipated." *Midwest Studies in Philosophy* 9:169–175.

——— 1987a. *The Latest on the Best: Essays on Evolution and Optimality.* Cambridge, Mass.: Bradford Books/MIT Press.

——— 1987b. "Human Kinds." In Dupré, 1987a.

——— 1990a. "Global versus Local Perspectives on Sexual Difference." In Rhode, 1990.

——— 1990b. "Probabilistic Causality: A Rejoinder to Ellery Eells." *Philosophy of Science* 57:690–698.

——— 1990c. "The Mental Lives of Nonhuman Animals." In *Interpretation and Explanation in the Study of Animal Behavior,* vol. 1, ed. M. Bekoff and D. Jamieson. Boulder, Colo.: Westview Press.

——— 1990d. "Scientific Pluralism and the Plurality of the Sciences: Comments on David Hull's *Science as a Process.*" *Philosophical Studies* 60:61–76.

Dupré, J., and N. Cartwright. 1988. "Probability and Causality: Why Hume and Indeterminism Don't Mix." *Noûs* 22:521–536.

Dwyer, D., and J. Bruce. 1988. *A Home Divided: Women and Income in the Third World.* Stanford: Stanford University Press.

Earman, J. 1986. *A Primer on Determinism.* Dordrecht: D. Reidel.

Eells, E. 1987. "Probabilistic Causality: Reply to John Dupré." *Philosophy of Science* 53:52–64.

——— 1988. "Probabilistic Causal Levels." In *Causation, Chance, Credence,* ed. B. Skyrms and W. Harper. Dordrecht: Kluwer Academic.

Eells, E., and E. Sober. 1983. "Probabilistic Causality and the Question of Transitivity." *Philosophy of Science* 50:35–57.

294   Bibliography

Fausto-Sterling, A. 1985. *Myths of Gender.* New York: Basic Books.
Feinberg, G. 1966. "Physics and the Thales Problem." *Journal of Philosophy* 63:5–17.
Feyerabend, P. 1963. "Mental Events and the Brain." *Journal of Philosophy* 90:295–296.
——— 1975. *Against Method.* London: New Left Books.
——— 1978. *Science in a Free Society.* London: New Left Books.
Fodor, J. 1968. *Psychological Explanation.* New York: Random House.
Foucault, M. 1978. *The History of Sexuality,* vol. 1, trans. R. Hurley. New York: Random House.
Galison, P. 1987. *How Experiments End.* Chicago: University of Chicago Press.
Gendron, B. 1970. "On the Relation of Neurological and Psychological Theories: A Critique of the Hardware Thesis." In *Boston Studies in the Philosophy of Science,* vol. 8, ed. R. C. Buck and R. S. Cohen. Dordrecht: D. Reidel.
Ghiselin, M. T. 1966. "On Psychologism in the Logic of Taxonomic Controversies." *Systematic Zoology* 15:207–215.
——— 1974. "A Radical Solution to the Species Problem." *Systematic Zoology* 23:536–544.
——— 1987. "Species Concepts, Individuality, and Objectivity." *Biology and Philosophy* 2:127–143.
Giere, R. 1984. *Understanding Scientific Reasoning.* 2d ed. New York: Holt, Rinehart and Winston.
Gifford, F. 1986. "Sober's Use of Unanimity in the Units of Selection Problem." In *PSA 1986,* vol. 1, ed. P. Asquith and P. Kitcher. East Lansing, Mich.: Philosophy of Science Association.
Glymour, C. 1980. *Theory and Evidence.* Princeton: Princeton University Press.
Goldberg, S. 1973. *The Inevitability of Patriarchy.* New York: William Morrow.
Goodman, N. 1965. *Fact, Fiction, and Forecast.* 2d ed. Indianapolis: Bobbs-Merrill.
Goosens, W. K. 1978. "Reduction by Molecular Genetics." *Philosophy of Science* 45:73–95.
Gould, S. J., and R. C. Lewontin. 1979. "The Spandrels of San Marco and the Panglossian Paradigm: A Critique of the Adaptationist Programme." *Proceedings of the Royal Society of London* 205:581–598.
Hacking, I. 1983. *Representing and Intervening.* Cambridge: Cambridge University Press.
——— Forthcoming. "The Disunities of Science." In *The Disunity of*

Science, ed. P. Galison and D. Stump. Stanford: Stanford University Press.

Hampshire, S. 1991. "Biology, Machines, and Humanity." In *The Boundaries of Humanity,* ed. J. J. Sheehan and M. Sosna. Berkeley: University of California Press.

Haraway, D. J. 1981–82. "The High Cost of Information in Post–World War II Evolutionary Biology: Ergonomics, Semiotics, and the Sociobiology of Communication Systems." *Philosophical Forum* 13:244–278.

—— 1985. "A Manifesto for Cyborgs: Science, Technology, and Socialist Feminism in the 1980s." *Socialist Review* 15:65–108.

—— 1989. *Primate Visions.* New York: Routledge, Chapman, and Hall.

Hare, R. M. 1952. *The Language of Morals.* Oxford: Oxford University Press.

Harré, R., and E. H. Madden. 1975. *Causal Powers.* Oxford: Blackwell.

Healey, R. 1978–79. "Physicalist Imperialism." *Proceedings of the Aristotelian Society* 79:191–211.

Hellman, G. P., and F. W. Thompson. 1975. "Physicalism, Ontology, Determination, and Reduction." *Journal of Philosophy* 72:551–564.

Hempel, C. G. 1965. *Aspects of Scientific Explanation.* New York: Free Press.

—— 1966. *Philosophy of Natural Science.* Englewood Cliffs, N.J.: Prentice-Hall.

Hennig, W. 1966. *Phylogenetic Systematics,* trans. D. D. Davis and R. Zangerl. Urbana: University of Illinois Press.

Hesse, M. 1980. "Theory and Observation." In *Revolutions and Reconstructions in the Philosophy of Science.* Brighton: Harvester Press.

Hesslow, G. 1976. "Two Notes on the Probabilistic Approach to Causality." *Philosophy of Science* 43:290–292.

Holmstrom, N. 1982. "Do Women Have a Distinct Nature?" *Philosophical Forum* 14:25–42.

Holton, G. 1973. *Thematic Origins of Modern Science.* Cambridge, Mass.: Harvard University Press.

Hornsby, J. 1980–81. "Which Physical Events Are Mental Events?" *Proceedings of the Aristotelian Society* 81:73–92.

Hrdy, S. 1981. *The Woman Who Never Evolved.* Cambridge: Cambridge University Press.

Hull, D. L. 1965. "The Effect of Essentialism on Taxonomy: 2000 Years of Stasis." *British Journal of Philosophy of Science* 15:314–326; 16:1–18.

—— 1972. "Reduction in Genetics—Biology or Philosophy." *Philosophy of Science* 39:491–499.

────── 1974. *The Philosophy of Biological Science*. Englewood Cliffs, N.J.: Prentice-Hall.

────── 1976. "Are Species Really Individuals?" *Systematic Zoology* 25:174–191.

────── 1988. *Science as a Process*. Chicago: University of Chicago Press.

────── 1989. *The Metaphysics of Evolution*. Albany, N.Y.: SUNY Press.

Hume, D. 1748. *An Enquiry concerning Human Understanding*. Reprint, Indianapolis: Hackett, 1977.

Jaggar, A. 1983. *Feminist Politics and Human Nature*. To-to-wa, N.J.: Rowman and Allanheld.

Jeffrey, R. 1971. "Statistical Explanation vs. Statistical Inference." In Salmon, Jeffrey, and Greeno, 1971.

Kant, I. 1785/1948. *The Moral Law* (*Groundwork of the Metaphysics of Morals*), ed. H. J. Paton. London: Hutchinson.

Kemeny, J. G., and P. Oppenheim. 1956. "On Reduction." *Philosophical Studies* 7:6–19.

Kim, J. 1978. "Supervenience and Nomological Incommensurables." *American Philosophical Quarterly* 15:149–156.

Kitcher, P. 1981. "Explanatory Unification." *Philosophy of Science* 48:507–531.

────── 1982. "Genes." *British Journal for the Philosophy of Science* 33:337–359.

────── 1984a. "Species." *Philosophy of Science* 51:308–333.

────── 1984b. "Against the Monism of the Moment." *Philosophy of Science* 51:616–630.

────── 1984c. "1953 And All That: A Tale of Two Sciences." *Philosophical Review* 93:335–376.

────── 1985. *Vaulting Ambition*. Cambridge, Mass.: MIT Press.

────── 1989. "Some Puzzles about Species." In *What the Philosophy of Biology Is*, ed. M. Ruse. Dordrecht: Kluwer Academic.

Kitcher, P., K. Sterelny, and C. K. Waters. 1990. "The Illusory Riches of Sober's Monism." *Journal of Philosophy* 87:158–161.

Kitts, D. B. 1983. "Can Baptism Alone Save a Species?" *Systematic Zoology* 32:27–33.

Kitts, D. B., and D. J. Kitts. 1979. "Biological Species as Natural Kinds." *Philosophy of Science* 46:613–622.

Kripke, S. 1972. "Naming and Necessity." In *Semantics of Natural Language*, ed. D. Davidson and G. Harman. Dordrecht: D. Reidel.

Kuhn, T. S. 1970. *The Structure of Scientific Revolutions*. 2d ed. Chicago: University of Chicago Press.

Lakatos, I. 1978. *The Methodology of Scientific Research Programmes,* ed. J. Worrall and G. Currie. Cambridge: Cambridge University Press.

Lewis, D. K. 1966. "An Argument for the Identity Theory." *Journal of Philosophy* 63:17–25.

——— 1973. "Causation." *Journal of Philosophy* 70:556–567.

Lewontin, R. C. 1974. *The Genetic Basis of Evolutionary Change.* New York: Columbia University Press.

Lewontin, R. C., and M. J. D. White. 1960. "Interaction between Inversion Polymorphisms of Two Chromosome Pairs in the Grasshopper, *Moraba scurra.*" *Evolution* 14:116–129.

Lewontin, R. C., S. Rose, and L. J. Kamin. 1984. *Not in Our Genes.* New York: Pantheon Books.

Lloyd, E. A. 1988. *The Structure and Confirmation of Evolutionary Theory.* Westport, Conn.: Greenwood Press.

Locke, J. 1689/1975. *An Essay concerning Human Understanding,* ed. P. H. Nidditch. Oxford: Oxford University Press.

Longino, H. 1990. *Science as Social Knowledge.* Princeton: Princeton University Press.

Mackie, J. L. 1974. *The Cement of the Universe.* Oxford: Oxford University Press.

Maull, N. L. 1977. "Unifying Science without Reduction." *Studies in the History and Philosophy of Science* 8:143–162.

May, R. M. 1973. *Stability and Complexity in Model Ecosystems.* Princeton: Princeton University Press.

Maynard-Smith, J. 1978. *The Evolution of Sex.* Cambridge: Cambridge University Press.

——— 1987. "How to Model Evolution." In Dupré, 1987a.

Mayr, E. 1945. *Birds of the Southwest Pacific.* New York: Macmillan.

——— 1963. *Animal Species and Evolution.* Cambridge, Mass.: Harvard University Press.

——— 1970. *Populations, Species, and Evolution.* Cambridge, Mass.: Harvard University Press.

——— 1987. "The Ontological Status of Species: Scientific Progress and Philosophical Terminology." *Biology and Philosophy* 2:145–166.

Mellor, D. H. 1977. "Natural Kinds." *British Journal for the Philosophy of Science* 28:299–312.

Mill, J. S. 1875. *System of Logic.* 8th ed. London: Longmans.

Millikan, R. G. 1984. *Language, Thought, and Other Biological Categories.* Cambridge, Mass.: MIT Press.

Mishler, B. D., and R. N. Brandon. 1987. "Individuality, Pluralism, and the Phylogenetic Species Concept." *Biology and Philosophy* 2:397–414.

Mishler, B. D., and M. J. Donoghue. 1982. "Species Concepts: A Case for Pluralism." *Systematic Zoology* 31:491–503.

Moore, G. E. 1922. "The Concept of Intrinsic Value." In *Philosophical Studies*. New York: Harcourt, Brace.

Morgan, E. 1972. *The Descent of Woman*. London: Corgi Books.

Nagel, E. 1961. *The Structure of Science: Problems in the Logic of Scientific Explanation*. New York: Harcourt, Brace and World.

Nelson, G., and N. Platnick. 1981. *Systematics and Biogeography*. New York: Columbia University Press.

Nickles, T. 1973. "Two Concepts of Intertheoretic Reduction." *Journal of Philosophy* 70:181–220.

Odling-Smee, F. J. 1988. "Niche-Constructing Phenotypes." In *The Role of Behavior in Evolution*, ed. H. C. Plotkin. Cambridge, Mass.: Bradford Books/MIT Press.

Odum, E. 1971. *Fundamentals of Ecology*. 3d ed. Philadelphia: Saunders.

Oppenheim, P., and H. Putnam. 1958. "The Unity of Science as a Working Hypothesis." In *Minnesota Studies in the Philosophy of Science*, vol. 2, ed. H. Feigl et al. Minneapolis: University of Minnesota Press.

Ortner, S., and H. Whitehead. 1981. *Sexual Meanings*. Cambridge: Cambridge University Press.

Paterson, H. E. H. 1985. "The Recognition Concept of Species." In *Species and Speciation*, ed. E. Vrba. Transvaal Museum Monograph No. 4. Pretoria: Transvaal Museum.

Peacocke, C. 1979. *Holistic Explanation*. Oxford: Oxford University Press.

Peirce, C. S. 1972. *The Essential Writings*, ed. E. C. Moore. New York: Harper & Row.

Place, U. T. 1956. "Is Consciousness a Brain Process?" *British Journal of Psychology* 47:44–50.

Popper, K. R. 1959. *The Logic of Scientific Discovery*. London: Hutchinson.

——— 1963. *Conjectures and Refutations: The Growth of Scientific Knowledge*. London: Routledge and Kegan Paul.

——— 1974. "Scientific Reduction and the Essential Incompleteness of All Science." In *Studies in the Philosophy of Biology*, ed. F. Ayala and T. Dobzhansky. Berkeley: University of California Press.

——— 1982. *The Open Universe*. Totowa, N.J.: Rowman and Littlefield.

Putnam, H. 1960. "Minds and Machines." In *Dimensions of Mind*, ed. S. Hook. New York: New York University Press.

——— 1963. "Brains and Behavior." In *Analytical Philosophy, Second Series*, ed. R. Butler. Oxford: Blackwell.

———— 1975. *Mind, Language, and Reality. Philosophical Papers*, vol. 2. Cambridge: Cambridge University Press.

———— 1978. *Meaning and the Moral Sciences*. London: Routledge and Kegan Paul.

———— 1981. *Reason, Truth and History*. Cambridge: Cambridge University Press.

Quine, W. V. O. 1951. "Two Dogmas of Empiricism," *Philosophical Review* 60. Reprinted in *From a Logical Point of View*. Cambridge, Mass.: Harvard University Press, 1953.

———— 1960. *Word and Object*. Cambridge, Mass.: MIT Press.

———— 1969. "Natural Kinds." In *Ontological Relativity and Other Essays*. New York: Columbia University Press.

Railton, P. 1978. "A Deductive-Nomological Model of Scientific Explanation." *Philosophy of Science* 45:206–226.

Rhode, D. 1990. *Theoretical Perspectives on Sexual Difference*. New Haven: Yale University Press.

Richerson, P., and R. Boyd. 1987. "Simple Models of Complex Phenomena: The Case of Cultural Evolution." In Dupré, 1987a.

Rorty, R. 1965. "Mind-Body Identity, Privacy, and Categories." *Review of Metaphysics* 19:24–54.

———— 1970. "In Defense of Eliminative Materialism." *Review of Metaphysics* 24:112–121.

Rosen, D. 1978. "Discussion: In Defense of a Probabilistic Theory of Causality." *Philosophy of Science* 45:604–613.

———— 1982–83. "A Critique of Deterministic Causality." *Philosophical Forum* 14:101–130.

Rosenberg, A. 1985. *The Structure of Biological Science*. Cambridge: Cambridge University Press.

———— 1989. "From Reductionism to Instrumentalism?" In *What the Philosophy of Biology Is*, ed. M. Ruse. Dordrecht: Kluwer Academic.

Rouse, J. 1987. *Knowledge and Power*. Ithaca, N.Y.: Cornell University Press.

Ruse, M. 1969. "Definitions of Species in Biology." *British Journal for the Philosophy of Science* 20:97–119.

———— 1971. "Reduction, Replacement, and Molecular Biology." *Dialectica* 25:39–72.

———— 1973. *The Philosophy of Biology*. London: Hutchinson.

———— 1976. "Reduction in Genetics." In *PSA 1974*, ed. R. S. Cohen et al. Dordrecht: D. Reidel.

Ruse, M., and E. O. Wilson. 1986. "Moral Philosophy as Applied Science." *Philosophy* 61:173–192.

Russell, B. 1953. "On the Notion of Cause with Applications to the Free-Will Problem." In *Readings in the Philosophy of Science,* ed. H. Feigl and M. Brodbeck. New York: Appleton-Century-Crofts.

Ryle, G. 1949. *The Concept of Mind.* London: Hutchinson.

Salmon, N. 1981. *Reference and Essence.* Princeton: Princeton University Press.

Salmon, W. C. 1980. "Probabilistic Causality." *Pacific Philosophical Quarterly* 61:50–74.

——— 1984. *Scientific Explanation and the Causal Structure of the World.* Princeton: Princeton University Press.

Salmon, W. C., R. C. Jeffrey, and J. G. Greeno. 1971. *Statistical Explanation and Statistical Relevance.* Pittsburgh: University of Pittsburgh Press.

Schaffner, K. F. 1967. "Approaches to Reduction." *Philosophy of Science* 34:137–157.

——— 1976. "Reductionism in Biology: Prospects and Problems." In *PSA 1974,* ed. R. S. Cohen et al. Dordrecht: D. Reidel.

Schlesinger, G. 1963. *Method in the Physical Sciences.* New York: Humanities Press.

Scriven, M. 1959. "Explanation and Prediction in Evolutionary Theory." *Science* 130:477–482.

Shapin, S., and S. Schaffer. 1985. *Leviathan and the Air-Pump.* Princeton: Princeton University Press.

Skyrms, B. 1980. *Causal Necessity.* New Haven: Yale University Press.

Smart, J. J. C. 1959. "Sensations and Brain Processes." *Philosophical Review* 68:141–156.

——— 1963. *Philosophy and Scientific Realism.* London: Routledge and Kegan Paul.

——— 1978. "The Content of Physicalism." *Philosophical Quarterly* 28:339–341.

Sneath, P. H. A., and R. R. Sokal. 1973. *Numerical Taxonomy.* San Francisco: W. H. Freeman.

Sober, E. 1984a. *The Nature of Selection.* Cambridge, Mass.: Bradford Books/MIT Press.

——— 1984b. "Sets, Species, and Evolution: Comments on Philip Kitcher's 'Species.'" *Philosophy of Science* 51:334–341.

——— 1990. "The Poverty of Pluralism: A Reply to Sterelny and Kitcher." *Journal of Philosophy* 87:151–158.

Sober, E., and R. C. Lewontin. 1982. "Artifact, Cause, and Genic Selection." *Philosophy of Science* 49:157–180.

Sokal, R. R., and P. H. A. Sneath. 1961. *Principles of Numerical Taxonomy.* San Francisco: W. H. Freeman.

Spellenberg, R. 1979. *The Audubon Society Field Guide to North American Wildflowers, Western Region.* New York: Alfred A. Knopf.

Sterelny, K., and P. Kitcher. 1988. "The Return of the Gene." *Journal of Philosophy* 85:339–361.

Stich, S. 1983. *From Folk Psychology to Cognitive Science.* Cambridge, Mass.: Bradford Books/MIT Press.

Suppes, P. 1970. *A Probabilistic Theory of Causality.* Amsterdam: North-Holland.

——— 1984. *Probabilistic Metaphysics.* Oxford: Blackwell.

Thompson, R. P. 1988. *The Structure of Biological Theories.* Albany: SUNY Press.

Thornhill, R., and N. Thornhill. 1983. "Human Rape: An Evolutionary Analysis." *Ethology and Sociobiology* 4:137–173.

Toulmin, S. 1972. *Human Understanding.* Princeton: Princeton University Press.

Traweek, S. 1988. *Beamtimes and Lifetimes.* Cambridge, Mass.: Harvard University Press.

van Fraassen, B. C. 1980. *The Scientific Image.* Oxford: Oxford University Press.

Van Valen, L. 1976. "Ecological Species, Multispecies, and Oaks." *Taxon* 25:233–239.

Waring, M. 1988. *If Women Counted.* San Francisco: Harper & Row.

Waters, C. K. 1990. "Why the Anti-Reductionist Consensus Won't Survive: The Case of Classical Mendelian Genetics." In *PSA 1990,* ed. A. Fine, M. Forbes, and L. Wessels. East Lansing, Mich.: Philosophy of Science Association.

Wiley, E. O. 1979. "An Annotated Linnaean Hierarchy, with Comments on Natural Taxa and Competing Systems." *Systematic Zoology* 28:308–337.

Williams, G. C. 1966. *Adaptation and Natural Selection.* Princeton: Princeton University Press.

——— 1975. *Sex and Evolution.* Princeton: Princeton University Press.

Wilson, E. O. 1975. *Sociobiology: The New Synthesis.* Cambridge, Mass.: Harvard University Press.

——— 1978. *On Human Nature.* Cambridge, Mass.: Harvard University Press.

Wimsatt, W. C. 1976. "Reductive Explanation: A Functional Account." In *PSA 1974,* ed. L. J. Cohen et al. Dordrecht: D. Reidel.

———— 1980. "Reductionist Research Strategies and Their Biases in the Units of Selection Controversy." In *Scientific Discoveries: Case Studies,* ed. T. Nickles. Dordrecht: D. Reidel.

Wittgenstein, L. 1953. *Philosophical Investigations.* Oxford: Blackwell.

Woodward, J. Forthcoming. "Capacities and Invariance." In *Philosophical Problems of the Internal and External Worlds: Essays concerning the Philosophy of Adolph Grunbaum,* ed. J. Earman et al. Pittsburgh: University of Pittsburgh Press.

Wright. L. 1976. *Teleological Explanations.* Berkeley: University of California Press.

Wylie, A. 1992. "The Interplay of Evidential Constraints and Political Interests: Recent Archaeological Research on Gender." *American Antiquity* 57:15–35.

Zymach, E. 1976. "Putnam's Theory on the Reference of Substance Terms." *Journal of Philosophy* 73:116–127.

Sources

More than half of the material in this book is entirely new; the rest has been revised to some degree. Chapters 1, 3, and 5 are substantially similar to previously published material. Significant parts of the following articles have been incorporated in the book. I extend thanks to the publishers of the journals and books in which they first appeared for permission to use them here.

"Natural Kinds and Biological Taxa," *Philosophical Review* 90 (1981), 66–91 (Chapters 1 and 2); copyright 1981 Cornell University

"Species," in *Keywords in Evolutionary Biology,* ed. E. F. Keller and E. Lloyd (Cambridge, Mass.: Harvard University Press, 1992) (Chapter 2)

"Sex, Gender, and Essence," *Midwest Studies in Philosophy* 11 (1986), 441–457 (Chapter 3)

"Materialism, Physicalism, and Scientism," *Philosophical Topics* 16 (1988), 31–56 (Chapters 4 and 7)

"The Disunity of Science," *Mind* 92 (July 1983), 321–346 (Chapters 4, 5, and 6)

"Probabilistic Causality Emancipated," *Midwest Studies in Philosophy* 9 (1984), 169–175 (Chapter 9)

"Probabilistic Causality: A Rejoinder to Ellery Eells," *Philosophy of Science* 57 (1990), 690–698 (Chapter 9)

"Scientific Pluralism and the Plurality of the Sciences: Comments on David Hull's *Science as a Process,*" *Philosophical Studies* 60 (1990), 61–76 (Chapter 10); reprinted by permission of Kluwer Academic Publishers

Index